T0362414

Renewable Energy and Green Technology

Renewable Energy and Green Technology

Principles and Practices

Edited by
Narendra Kumar, Hukum Singh and
Amit Kumar

CRC Press
Taylor & Francis Group
Boca Raton London New York

CRC Press is an imprint of the
Taylor & Francis Group, an **informa** business

First edition published 2022
by CRC Press
6000 Broken Sound Parkway NW, Suite 300, Boca Raton, FL 33487-2742

and by CRC Press
2 Park Square, Milton Park, Abingdon, Oxon, OX14 4RN

Library of Congress Cataloging-in-Publication Data
Names: Kumar, Narendra, 1976–, editor. | Singh, Hukum, 1983– editor. |
Kumar, Amit (Assistant professor in biotechnology), editor.
Title: Renewable energy and green technology: principles and practices /
edited by Narendra Kumar, Hukum Singh and Amit Kumar.
Description: First edition. | Boca Raton: CRC Press, 2022. |
Includes bibliographical references and index.
Identifiers: LCCN 2021031727 (print) | LCCN 2021031728 (ebook) |
ISBN 9781032008158 (hbk.) | ISBN 9781032008165 (pbk.) | ISBN 9781003175926 (ebk.)
Subjects: LCSH: Renewable energy sources. | Green technology.
Classification: LCC TJ808 .R46 2022 (print) | LCC TJ808 (ebook) | DDC 621.042—dc23
LC record available at https://lccn.loc.gov/2021031727
LC ebook record available at https://lccn.loc.gov/2021031728

ISBN: 9781032008158 (hbk)
ISBN: 9781032008165 (pbk)
ISBN: 9781003175926 (ebk)

DOI: 10.1201/9781003175926

Typeset in Times
by codeMantra

Contents

Editors

Dr. Narendra Kumar earned his PhD in Plant Physiology from G.B. Pant University of Agriculture and Technology, Pantnagar, India. His research work is focused on the response of plants to abiotic stresses. Later, he served as Assistant Professor in various reputed academic institutes, especially Surajmal Agarwal (Pvt.) Kanya Mahavidyalaya, Kichha US Nagar; Dev Bhoomi Group of Institutes, Dehradun, India; and Uttaranchal University, Dehradun, India. He has published various research articles, book chapters, and articles in peer-reviewed journals. Currently, he is working as Research Associate at Forest Ecology and Climate Change Division, Forest Research Institute, Dehradun, India.

Dr. Hukum Singh, Scientist at Forest Ecology and Climate Change Division, Forest Research Institute, Dehradun – India, has been actively involved in research and teaching. He has received several awards, i.e. SERB-DST fast-track young scientist research award, UCOST young scientist award, AsiaFlux Japan travel fellowship, VIFRA young scientist research award, and VIFA outstanding faculty award. He is a member of various national and international scientific societies/organizations. His research areas focus primarily on ecophysiology, plant response to abiotic stresses, green technology, climate change adaptation and mitigation functional traits, carbon-water-energy flux, and climate change modeling. His previous research was funded by ICFRE, SERB-DST, DST, ISRO, NMHS, MoEF&CC, and SFDs. He has supervised many MSc and PhD students and post-doctoral Fellows and many more are under supervision. He has published various papers/chapters/articles in national and international peer-reviewed reputed journals with high impact factors.

Mr. Amit Kumar is pursuing his PhD in Forestry at Forest Research Institute (Deemed to be University), Dehradun, India. His research work is mainly focused on characterizing plant functional traits of forestry species. Earlier, he worked as Assistant Professor in many organizations, i.e. Dolphin P.G. Institute of Biomedical and Natural Sciences, Dehradun and Alpine Group of Institutions, Dehradun, Uttarakhand. He has published several research articles and book chapters in peer-reviewed journals.

Contributors

Amit Bijlwan
G.B. Pant University of Agriculture and
    Technology
Pantnagar, India

Monika Bisht
Division of Natural Resource Management
Doon University
Dehradun, India

Sarita Bisht
Forest Ecology and Climate Change Division
Forest Research Institute
Dehradun, India

Aditee Das
Tezpur University
Sonitpur, India

Roop Jyoti Das
Forest Research Institute
Dehradun, India

Gagan Dixit
Department of Physics
G.B. Pant University of Agriculture and
    Technology
Pantnagar, India

Deepika Gabba
Department of Microbiology, Akal College of
    Basic Sciences
Eternal University
Sirmaur, India

Krishna Giri
Rain Forest Research Institute
Jorhat, India

Upasana Gola
Department of Microbiology, Akal College of
    Basic Sciences
Eternal University
Sirmaur, India

Neha Jeena
Department of Biotechnology,
    Bhimtal Campus
Kumaun University
Nainital, India

D. Jhajharia
Soil and Water Conservation Engineering
    Department, College of Agricultural
    Engineering & Post Harvest Technology
Central Agricultural University
Imphal, India

Chandra Kanta
Department of Botany
Doon (PG.) College of Agriculture, Sciences and
    Technology
Dehradun, India

Rowndel Khwairakpam
School of Agriculture,
Graphic Era Hill University
Dehradun, India

Parmanand Kumar
Forest Ecology and Climate Change Division
Forest Research Institute
Dehradun, India

Garima Kumari
Forest Ecology and Climate Change Division
Forest Research Institute
Dehradun, India

Puneet Negi
Department of Physics, Akal College of Basic
    Sciences
Eternal University
Sirmaur, India

Pankaj Singh Rawat
Department of Physics
G.B. Pant University of Agriculture and
    Technology
Pantnagar, India

Sunita Rawat
Forest Genetics and Tree Improvement Division
Forest Research Institute
Dehradun, India

Meenakshi Sati
Forest Research Institute
Deemed to be University
Dehradun, India

Ishwar Prakash Sharma
Patanjali Herbal Research Department
Patanjali Research Institute
Haridwar, India.

Dipti Singh
Division of Genetics
Indian Agricultural Research Institute
New Delhi, India

Manali Singh
Invertis Institute of Engineering and Technology
  (IIET)
Invertis University
Bareilly, India

Manendra Singh
Genetics and Tree Improvement Division
Forest Research Institute
Dehradun, India

Nasib Singh
Department of Microbiology, Akal College of
  Basic Sciences
Eternal University
Sirmaur, India

Rajat Singh
Forest Ecology and Climate Change Division
Forest Research Institute
Dehradun, India

Ravindra Soni
Department of Agricultural Microbiology,
  College of Agriculture
Indira Gandhi Krishi Vishwa Vidyalaya
Raipur, India

R.C. Srivastava
Department of Physics
G.B. Pant University of Agriculture and
  Technology
Pantnagar, India

Deep Chandra Suyal
Department of Microbiology, Akal College of
  Basic Sciences
Eternal University
Sirmaur, India

Neha Suyal
Akal College of Nursing
Eternal University
Sirmaur, India

Anugrah Tripathi
Indian Council of Forestry Research and
  Education
Dehradun, India

Shivani Uniyal
Department of Botany
Banaras Hindu University
Varanasi, India

Megha Verma
Doon (PG) College of Agriculture Science and
  Technology
Dehradun, India

Ratnakiran D. Wankhade
G.B. Pant University of Agriculture and
  Technology
Pantnagar, India

Shambhavi Yadav
Genetics and tree improvement Division
Forest Research Institute
Dehradun, India

G.S. Yurembam
Soil and Water Conservation Engineering
  Department
College of Agricultural Engineering & Post
  Harvest Technology
Central Agricultural University
Imphal, India

1

Fundamentals of Energy: Its Potentials and Achievements

Meenakshi Sati and Megha Verma

CONTENTS

1.1 Introduction

Energy can be defined in many ways; however, energy from the sun is the ultimate source of energy for all living things. In physics, energy is defined as the capacity to do work, which is measured as the potential energy or the kinetic energy. Contrarily, in economics, energy is regarded as fuel (Martinás, 2005). Energy is valuable only when it is effortlessly convertible to useful work, and fossil fuels account for most of the world's convertible energy, thereby playing an important part in affecting human civilization. Sun is the primary source of energy on earth providing all living things, including our bodies, the capability to live, grow, repair tissue, reproduce, and do work. As it does so, energy changes from one form to another, i.e., from electromagnetic radiation flowing through space to chemical energy stored in plants, to heat that keeps us warm, to potential energy as we climb a mountain, and to kinetic energy as we move back down again. In order to heat our houses, produce our electricity, and drive our vehicles,

DOI: 10.1201/9781003175926-1

we use solar energy that plants gathered and stored long ago. It is continuously converted from one form to another as the energy works its way through nature and human societies. Though none is lost as it is processed, transformed, and used, its content is continuously reduced to less helpful types, ultimately ending up as low-temperature heat that is relatively useless.

The human civilization is illustrated in terms of new sources or forms of energy used or developed in that era, e.g., fossil fuel era, post-fossil fuel era, nuclear era, an era of renewable, etc. Industrial revolutions in history were energy revolutions. The first industrial revolution (1760–1840) witnessed the beginning of mechanization that led to the formation of a new economic structure. Coal-powered steam engines brought about revolutions in the production system and communication network; handlooms were substituted by power looms; railway and steamboats minimized the distance. Amid the appearance of a new energy source, i.e., electricity, gas, and oil, the second industrial revolution started in the late 19th century (1870) and extended up to World War I. This was fundamentally a technological revolution indicated by the electrification of manufacturing systems that made possible the assembly line and mass production. In 1969, the third industrial revolution started with the commencement of nuclear power generation. A remarkable development in electronics, telecommunication, and computing technology, especially programmable logic controllers and robots, led to the automation of the production system (Schwab, 2016). The fourth industrial revolution has already started integrating the production processes with alternative energy resources such as wind, sun, and geothermal energy (Sentryo, 2017).

1.2 Energy Sources

There are various energy sources all over the world. It is possible to identify conventional energy sources as non-renewable and renewable. Due to the very long period taken for them to be replenished in nature, non-renewable energy resources, such as coal, nuclear, oil, and natural gas, are restricted in reserve and availability. Renewable resources, on the other hand, are replenished indefinitely or over very short periods. Solar, wind, hydro, biomass, geothermal, and marine or ocean energy are major renewable energy options (Andrea, 2014). Some unconventional sources have emerged as energy sources in recent years, such as waste-to-energy (WtE), carbon capture, storage, etc.

1.3 Non-renewable Energy Sources

1.3.1 Coal

The converted remnants of ancient vegetation amassed in swamps and peat bogs are coal. It is an inflammable sedimentary organic rock consisting predominantly of carbon, hydrogen, and oxygen. Coals are categorized into peat, lignite, sub-bituminous, bituminous, anthracite, and graphite based on their physical and chemical characteristics formed during coalification. Coal accounts for a large proportion (around 30%) of the world's primary energy consumption in several industries, including power generation, iron and steel production, cement production, etc. Around 40% of global power plants are currently burning coal; 70% of steel production and 50% of aluminium production plants use coal, on the other hand (WCA, 2009). The mining and use of coal causes numerous premature deaths and various diseases. As the largest anthropogenic source of carbon dioxide, 14.4 gigatonnes (Gt) in 2018, which is 40% of total fossil fuel emissions and over 25% of total global greenhouse gas emissions, the coal industry harms the atmosphere, leading to climate change. Many countries have limited or discontinued their use of coal-fired power as part of the global energy transition by 2020. Coal usage peaked in 2013, but coal use needs to halve from 2020 to 2030 in order to meet the Paris Agreement goal of holding global warming well below 2°C (3.6°F). The largest consumer and importer of coal is China. China mines almost half the world's coal, followed by India at about a tenth. Australia accounts for about a third of world coal exports, followed by Indonesia and Russia (IEA, 2020c).

1.3.2 Oil

Petroleum, also called crude oil or just oil, is a naturally occurring, yellowish-black substance found under the surface of the earth in geological formations. It is commonly processed into different kinds of fuels. It consists of hydrocarbons of different molecular weights that occur naturally and can contain various organic compounds (DOE, 2017). Petroleum has been recovered mainly by oil exploration. In crude oil, the hydrocarbons are primarily alkanes, cycloalkanes, and various aromatic hydrocarbons. At the same time, the other organic compounds contain nitrogen, oxygen, and sulphur, as well as trace quantities of metals, such as iron, nickel, copper, and vanadium. Oil accounts for a significant percentage of energy consumption worldwide, ranging from a low of 32% for Europe and Asia to a high of 53% for the Middle East. The world consumes 36 billion barrels (5.8 km^3) of oil every year, with the greatest users being developed nations (Sönnichsen, 2020). In 2015, the United States consumed 18% of the oil produced (CIA, 2020). In terms of dollar value, the production, distribution, refining, and retailing of oil taken as a whole represents the world's largest industry.

1.3.3 Natural Gas

Natural gas is a mixture of naturally occurring hydrocarbon gas that consists mainly of methane but typically contains varying quantities of other higher alkanes, and often a small percentage of carbon dioxide, nitrogen, sulphide hydrogen, or helium (IEA, 2014). It is produced when decomposing layers of plant and animal matter are exposed to extreme heat and pressure (Markandya and Wilkinson, 2007). The combustion of natural gas is cleaner than that of other fuels, such as oil and coal. It produces less carbon dioxide per unit of energy emitted than coal, which produces mostly carbon dioxide since burning natural gas produces both water and carbon dioxide. Natural gas is a non-renewable hydrocarbon used for heating, cooking, and electricity generation as a source of energy. It is also used to manufacture plastics and other commercially relevant organic chemicals as fuel for automobiles and as a chemical feedstock. Natural gas is found in deep underground rock formations or is associated in coal beds and as methane clathrates with other hydrocarbon reservoirs. Another fossil fuel contained next to and with natural gas is petroleum. Two processes have produced the majority of natural gas over time: biogenic and thermogenic. In marshes, bogs, landfills, and shallow sediments, biogenic gas is produced by methanogenic organisms. Deep inside the earth, thermogenic gas is produced from buried organic material at higher temperatures and pressure (U.S. EPA, 2014; U.S. EIA, 2020).

Worldwide growth in natural gas consumption reached 2% in 2019, lower than its 10-year average and significantly down from the extraordinary growth seen in 2018 (5.3%). Demand has risen by 78 billion cubic meters (bcm) in volume terms, led by the US (27 bcm) and China (24 bcm). The share of gas in primary oil, however, rose to a record high of 24.2%. Production of gas increased by 132 bcm (3.4%), outpacing demand development. The US accounted for nearly two-thirds of global net expansion, with the 85 bcm volumetric rise just shy of the record increase in 2018 (90 bcm). In Australia (23 bcm) and China, supply was also boosted by strong growth (16 bcm). World proven gas reserves grew in 2019 by 1.7 Tcm to 198.8 Tcm. The biggest rises were made by China (2 Tcm) and Azerbaijan (0.7 Tcm), but this was partly offset by a 1.3 Tcm decrease in Indonesian reserves. The countries with the greatest reserves are Russia (38 Tcm), Iran (32 Tcm), and Qatar (24.7 Tcm) (BP plc. 2020; U.S. EIA, 2021).

1.3.4 Nuclear Energy

Nuclear power refers to the use of nuclear reactions that release nuclear power to produce heat, which is then used most commonly in steam turbines to generate electricity in a nuclear power plant. From nuclear fission to nuclear decay and nuclear fusion reactions, nuclear power can be produced. Currently, nuclear fission of uranium and plutonium generates the vast majority of energy from nuclear power. Compared to other energy sources, nuclear power has one of the lowest levels of fatalities per unit of energy produced (IAEA, 2021).

In 2019, civilian nuclear power supplied 2,586 terawatt-hours (TWh) of energy, equal to about 10% of global power generation, and was the second-largest source of low-carbon power after hydroelectricity (IEA, 2019; WNA, 2021). As of December 2020, the world has 441 civilian fission reactors with a total electrical capacity of 392 gigawatts (GW). There are also 54 nuclear power reactors under construction and 98 proposed reactors, with a total capacity of 60 GW and 103 GW, respectively (WNA, 2021). The United States has the largest nuclear reactor fleet, producing over 800 TWh of zero-emission electricity per year with an estimated capacity factor of 92% (DOE, 2020; IAEA, 2021).

1.4 Renewable Energy Sources

1.4.1 Hydropower

The power extracted from the energy of falling or fast-running water, which can be harnessed for useful purposes, is hydropower. Hydropower from many types of watermills has been used numerous times since ancient times as a source of renewable energy for irrigation and the operation of various mechanical devices, such as gristmills, sawmills, cloth mills, travel hammers, dock cranes, domestic lifts, and ore mills (DOE, 2017).

Hydropower became a source of electricity production in the late 19th century. The first hydroelectric powerhouse was built in 1878 in Cragside in Northumberland, and the first commercial hydroelectric power plant was built in 1879 at Niagara Falls (IHA, 2016). The term has been used almost entirely in connection with the new production of hydroelectric power since the early 20th century. Without contributing large quantities of pollution to the atmosphere, multinational organisations such as the World Bank perceive hydropower as a tool for economic growth. Still, dams can have major negative social and environmental impacts. As of 2019, traditional hydroelectric power plants with dams are the world's five largest power stations. Other types of hydropower electricity generation include tidal stream generators that use tidal power from oceans, rivers, and man-made canal systems to produce electricity. In the world, hydropower alone produces 71% of all renewable energy. China's Three Gorges Dam is the world's largest hydropower plant with a huge 22.5 GW capacity, generating an annual average of 88.2 TWh of electricity (WEC, 2016). Over the forecast period, global hydropower production (excluding pumped storage) is expected to increase by 9.5%, to grow from 4,250 TWh in 2019 to 4,650 TWh in 2025, and to remain the world's largest source of renewable electricity (IEA, 2020c).

1.4.2 Bioenergy

Bioenergy is biomass or biofuel-based energy. Biomass is any organic material that has absorbed sunlight and stored it in the form of chemical energy. The main source of biomass energy today is wood and wood waste. Wood may be directly used as a fuel or converted into pellet fuel or other types of fuel. Other plants, such as maize, switchgrass, miscanthus, and bamboo, can also be used as fuel. Wood waste, agricultural waste, municipal solid waste, and production waste are the major waste feedstocks. Different processes, commonly known as thermal, chemical, or biochemical, may be used to upgrade raw biomass to higher-quality fuels (Liu et al., 2011; Kumar et al., 2016; Akhtar et al., 2018).

The two sources of bioenergy supply are forest and agriculture, with woody biomass being a major primary source of energy worldwide. Contemporary bioenergy is an important source of renewable energy, and its contribution to final energy demand across all sectors is five times greater than the combination of wind and solar PV, even if biomass is excluded from traditional use (Kumar et al., 2020). Bioenergy for electricity and biofuels for transport has been growing rapidly in recent years, primarily due to higher levels of policy support. The heating sector, however, remains the largest bioenergy source. In developing countries and emerging economies, modern bioenergy does not include the traditional use of biomass for cooking and heating purposes, the use of inefficient open fires, or simple cooking stoves that have an impact on human health and the environment. Although biomass supplies about 59.2 EJ of energy at present, which is 10.3% of the global final energy use, it can provide as much as 150 EJ energy by 2035 (IRENA, 2014; Kumar et al., 2015).

Bioenergy supplies around 9% of the demand for industrial heat and is concentrated in bio-based sectors such as paper and board. Biofuels, mostly ethanol and biodiesel, account for about 3% of transport energy and 5% of global biofuel production increase in 2019. Despite a decline in the United States, the major ethanol producer, ethanol production grew around 2%. The production of biodiesel increased by 13%, and Indonesia became the largest producer globally, overtaking the United States, where production declined by 7%. The contribution of bioenergy in the electricity sector rose by 9% in 2019 to 501 terawatt-hours (TWh). As the country's largest producer, China expanded its lead, and bio-electricity development was also high in the EU, Japan, and Korea (REN21, 2020).

1.4.3 Solar Energy

Solar energy uses the Sun's radiant light and heat harnessed across various ever-evolving technologies such as solar heating, photovoltaics, solar thermal energy, solar architecture, molten salt power plants, and artificial photosynthesis (IEA, 2011). It is an important source of renewable energy, and its systems, depending on how they absorb and transmit solar energy or transform it into solar electricity, are broadly defined as either passive solar or active solar.

In order to turn sunlight into usable outputs, active solar techniques use photovoltaics, concentrating solar power, solar thermal collectors, pumps, and fans. The use of materials with desirable thermal properties, the construction of spaces that automatically flow air, and the relation to the orientation of a building to the Sun are passive solar strategies. Active solar technologies boost electricity availability and are perceived to be supply-side technologies. In contrast, passive solar technologies minimize the requirement for alternative fuels and are commonly seen as demand-side technologies (Philibert, 2005). In two primary ways, solar power is generated:

a. *Photovoltaics (PV):* also referred to as solar cells, are electrical devices that directly turn sunlight into electricity.

b. *Concentrated solar power (CSP):* employs mirrors to focus solar rays. These rays heat fluid, generating steam to drive and produce energy from a turbine. In large-scale power plants, CSP is used to produce electricity.

Photovoltaic development worldwide is highly diverse and varies strongly by region. A total sum of 629 GW of solar power was installed worldwide at the end of 2019 (IEA, 2020e). At 208 GW, China was the leading solar power country by the beginning of 2020, accounting for one-third of the world's installed solar energy (Statista, 2020; CT, 2020; IEA, 2020b). There are at least 37 countries with a combined PV capacity of more than 1 GW worldwide as of 2020. China, United States, and India were the top installers from 2016 to 2019 (IEA, 2020e).

1.4.4 Geothermal Energy

The thermal energy produced and deposited on the earth is geothermal energy. The force that determines the temperature of matter is thermal energy. The earth's crust's geothermal energy originates from the planet's initial creation and radioactive material decay. Geothermal power is cost-effective, efficient, sustainable, and environmentally friendly but has traditionally been restricted to areas close to the borders of the tectonic plates (Khan, 2007).

Over the 3 years to 2015, foreign markets expanded at an average annual rate of 5%, and global geothermal power capacity is projected to hit 14.5–17.6 GW by 2020 (GEA, 2015). In 2019, geothermal electricity production totalled around 95 TWh, while direct useful thermal output reached around 117 TWh (421 petajoules). In 26 countries, geothermal power generation is currently used, while geothermal heating is in 70 countries. Global geothermal power capacity amounts to 15.4

gigawatts (GW) as of 2019, out of which 23.86% or 3.68 GW is installed in the United States (Richter, A., 2020). The Geothermal Energy Association (GEA) reports that only 6.9% of the estimated global capacity has been exploited so far, based on existing geological information and technologies publicly disclosed by the GEA. In contrast, the IPCC has recorded geothermal power output in the range of 35 GW to 2 TW. El Salvador, Kenya, the Philippines, Iceland, New Zealand, and Costa Rica are countries that generate more than 15% of their electricity from geothermal sources (GEA, 2015).

As heat extraction is limited relative to the earth's heat content, geothermal power is known to be a reliable, renewable source of electricity. On average, the emissions of greenhouse gases from geothermal power stations are 45 g of carbon dioxide per kilowatt-hour of electricity, or less than 5% of that from conventional coal-fired power plants (Moomaw et al., 2011). Geothermal has the capacity to reach 3%–5% of global demand by 2050 as a form of clean electricity for both power and heating. It is projected, with economic benefits, that it will be possible to reach 10% of global demand by 2100 (Rybach, 2009).

1.4.5 Wind Energy

Wind energy is harnessed by wind turbines which spin generators, creating electricity. Wind power is a common source of sustainable, renewable energy that has a much lower environmental impact than the burning of fossil fuels. Wind farms consist of several individual wind turbines connected to the transmission network of electricity. With a worldwide installed wind power capacity of about 600 gigawatts, wind supplies about 5% of global electricity generation (CCES, 2020). In 2019, the worldwide wind power industry rose 19% to 60 GW, the second-highest annual growth, for a total of 650 GW (621 GW onshore and the rest offshore). The strong growth was primarily due to surges in China and the United States in the run-up to policy reforms and a substantial rise in Europe amid the continuing downturn of the economy in Germany. In 2019, wind energy accounted for an estimated 57% of Denmark's power production, with high shares in Ireland (32%), Uruguay (29.5%), Portugal (26.4%), and many other nations as well (REN21, 2020).

1.4.6 Ocean Energy

Ocean energy, also often referred to as marine energy or hydrokinetic energy, refers to energy from ocean waves, tides, salinity, and variations in ocean temperature. A large store of kinetic energy, or energy in motion, is generated by water movement in the world's oceans. It is possible to use some of this energy to produce electricity to fuel houses, transport, and factories. Both wave power, i.e., surface wave power, and tidal power, i.e., derived from the kinetic energy of large moving water bodies, are included in the term marine energy (CT, 2006). There is a large amount of energy in the oceans, and they are close to many, if not most, concentrated populations. Ocean energy can produce a large amount of new renewable energy worldwide. There is the potential to produce 20,-000–80,000 terawatt-hours of electricity per year (TWh/y) produced by changes in ocean temperatures, salt content, tidal movements, winds, waves, and swells (IEA, 2016). Ocean energy's resource capacity is immense, but the production direction of ocean power technologies has been volatile, and these capabilities remain largely untapped. The smallest part of the renewable energy industry has been ocean fuel, and the bulk of installations to date have been small-scale demonstration and pilot projects. In 2019, net additions were nearly 3 MW with a projected 535 MW of operating capacity at the end of the year. For 2020 and beyond, major investments and deployments have been scheduled. The growth of ocean power was mainly concentrated in Europe, where tidal stream devices generated 15 gigawatt-hours (GWh) in 2019. In Canada, the United States, and China, however, ocean power has gained momentum, providing generous revenue funding and innovative research and development (R&D) programs (REN21, 2020).

1.4.7 Waste-to-Energy

A considerable amount of energy can be recovered from MSW, construction waste, biowaste, medical waste, etc. depending on their specific composition and energy content. Waste-to-energy is the process of generating electricity and heat from the primary treatment of waste. Incineration of waste is used to produce steam in a boiler that drives a steam turbine to produce power. Japan uses up to 60% of its solid waste for incineration. The largest circulating fluidized bed waste-to-energy plant in China, which was built in 2012, can process 800 tonnes of waste per day. In China, at the beginning of 2016, there were about 434 waste-to-energy plants. With 40 million tons, Japan is the world's largest user of thermal treatment for urban solid waste (Fanchi and Fanchi, 2016). As an essential component of advanced waste management solutions, waste-to-energy is gaining status under which it plays the role of an alternate approach to alleviate landfill pressure. Waste-to-energy's additional value over other waste disposal techniques is the capacity for energy production. The plant itself uses a large portion of this electricity for its own energy requirements; the rest is provided to the city. As urbanization and industrial growth contribute to more local solid waste processing, waste-to-energy implementation is also rising strongly. For cities to handle urban solid waste, this technology provides a solution superior to landfills, and China has the largest waste-to-energy capability deployed globally. In 2019, the worldwide size of the waste-to-energy industry hit US$35.1 billion. The waste-to-energy market is projected to be US$50.1 billion by 2027, rising from 2020 to 2027 at a compound annual growth rate of 4.6% (Tiseo, I., 2020).

1.4.8 Bioenergy with Carbon Capture and Storage (BECCS)

BECCS is the mechanism by which bioenergy is derived from biomass, and the carbon is preserved and deposited, thereby eliminated from the environment (Obersteiner, 2001). The carbon in the biomass comes from the carbon dioxide (CO_2) greenhouse gas that the biomass absorbs from the atmosphere as it rises. As the biomass is used by combustion, fermentation, pyrolysis, or other conversion processes, energy is derived in useful forms (electricity, heat, biofuels, etc.). Some of the carbon in the wood is converted to CO_2 or biochar, which can then be stored through geological sequestration or land application, allowing the reduction of carbon dioxide and converting BECCS into technology for negative emissions (NASEM, 2019). The Intergovernmental Panel on Climate Change (IPCC) Fifth Assessment Report recommends a possible spectrum of BECCS negative emissions of 0–22 gigatonnes per year (Smith and Porter, 2018). Five plants around the world have been successfully using BECCS technology as of 2019 and have captured nearly 1.5 million tonnes of CO_2 each year (Consoli, 2019). The broad deployment of BECCS is limited by biomass costs and availability (Rhodes and Keith, 2008; Fajardy et al., 2019).

1.5 Energy Conservation

1.5.1 Energy Efficiency

Energy efficiency is required to combat climate change, clean the air, and save money. Via transmission, heat loss, and outdated infrastructure, electricity is wasted every year, costing money for families and companies and contributing to increased carbon emissions and climate change. Energy conservation is one of the fastest and most cost-effective forms of battling climate change, cleaning the environment, and saving money for customers and companies. To achieve greater energy efficiency, there are several choices. It is possible to achieve energy efficiency by:

- Usage of energy-efficient lighting, room cooling and heating systems, refrigeration, and so on.
- Better construction of our buildings and landscapes.
- Changing our acts.

- Improve device operations to optimize several functions inside a building, business, city, or other location, and, as a result, reduce energy consumption.
- It is important that we move our energy systems away from fossil fuels towards low-carbon energy sources to minimize global emissions.

1.5.2 Energy Management and Energy Policy

Energy management involves energy saving by introducing energy-efficiency practices and determining the type of energy and energy tariff suitable for the organization concerned, such as power stations, manufacturing plants, residential and commercial buildings, etc. The key goal of energy management is to ensure efficient procurement and usage of energy and reduce energy costs and reduce energy waste without impacting efficiency and quality, with minimal environmental impact. Energy management has become an unassailable part of the current industrial management system because of the connection between energy and the environment.

Energy policy is a mechanism in which the government (or any organization) deals with issues related to the growth and use of energy, including the production, distribution, and use of energy. Energy policy characteristics can include regulations, international treaties, investment incentives, energy efficiency advice, taxation, and other public policy techniques (Armstrong et al 2016). An energy policy is a working paper that states how an organization's energy will be handled. The following should be discussed in a systematic energy policy (CT, 2018):

- To ensure that energy services and a consistent supply are sufficiently available.
- To ensure compliance with legislation relating to energy and climate change.
- To ensure progress in cost-effective programs in the field of energy efficiency.
- To make the use of facilities and services that are energy-efficient and to manage energy costs.

1.6 Importance of Renewable Energy

The continued depletion of fossil fuels and the environmental challenges raised by potential growth needs are increasingly changing the development direction towards sustainability, enhanced sociability, and environmental responsibility, emphasizing the need for renewable sources of energy. The field of renewable energy sources is rising rapidly, and there are various developments as well as applications. The idea of decentralized renewable energy systems has been recognized as a solution to both the household and the agro-industrial sector's energy demands. Renewable energy is energy derived from natural resources that replenish themselves without depleting the resources of the earth in less than a human lifetime. These tools are available almost anywhere in one way or another, such as sunshine, wind, rain, tides, waves, biomass, and thermal energy deposited in the earth's crust. They are inexhaustible, and no environmental damage is caused by them.

On the other hand, fossil fuels, such as gasoline, coal, and natural gas, are only usable in finite amounts. They will run out sooner or later when we keep removing them. Natural resource depletion and increased demand for conventional resources have driven planners and policymakers to hunt for alternative sources. Even though commercial energy sources such as coal, oil, and natural gas are currently being used to a large degree, renewable energy sources are increasingly gaining importance. Renewable energy plays a fundamental role in sustainable development. These sources will provide the energy we need, polluting much less than fossil fuels for infinite periods. To increase diversity in energy supply markets, the benefits of renewables are well known; to secure long-term sustainable energy supplies; to minimize local and global atmospheric emissions; and to generate new jobs that provide opportunities for local development. Renewable energy has been rising higher than all other sources of energy since 2011. Another record-breaking year in 2019 was clean energy, as installed power capacity increased by more than 200 gigawatts (GW), the biggest growth ever (IEA, 2020a, d). Global increase in road transport, manufacturing activity, and power generation dependent on fossil fuel (as well as the free burning of waste in many cities) leads to high air pollution levels. The use of charcoal and fuelwood for heating and cooking in many developing nations also leads to low indoor air quality. Particles and other fossil-fuel air emissions asphyxiate towns. According to World Health Organization reports, their presence over urban sky is responsible for millions of premature deaths and costs billions. On the other hand, renewable energy releases no or low levels of air pollutants, which is safer for our welfare (REN21, 2020).

1.7 World Energy Consumption

The world consumes about 100 million barrels of oil a day at present. While the IEA aims to increase renewable energy, boost energy efficiency, and turn to electric vehicles, the oil will continue to meet the growing demand for petrochemicals and transport fuel. As countries aim to reduce greenhouse gas emissions by displacing coal for heating and power generation, natural gas demand will rise. Besides generating 40%–65% less pollution than coal, natural gas is cost-effective, plentiful, and secure. Who uses the most energy? The response depends on the energy type. Though China is the top user of electricity in the world, the United States consumes the most oil. Less energy is used by the average Chinese family than by the average North American family, but that is changing. As prosperity rises in countries such as China and India, and more people start living middle-class lifestyles, energy demand may increase further. Figure 1.1 depicts the total energy consumption worldwide (Enerdata, 2020). Different countries use various forms of energy, so it is hard to tell who uses the most energy. A complex energy mix of oil, natural gas, hydroelectricity, nuclear power, and more is used by Canada (IEA, 2020c).

Demand is rising regardless of the source of energy. With the global population projected to grow by around two billion over the next two decades and with living standards rising, it is predicted that 49% of electricity demand is expected to increase by 2040. Fossil fuels generate about 80% of the energy we require. Nuclear power, biofuels, hydropower, and other renewables, such as solar, wind, and geothermal, are the remaining sources. In its 2020 report, the IEA predicts that global demand for natural gas will increase by 29% by 2040, supply 25% of the world's total energy consumption, and increase global oil demand by 7%, providing 28% of the world's total energy consumption. The IEA's annual forecasts suggest that global energy demand will continue to increase as the world expands and reduces poverty (IEA, 2020c).

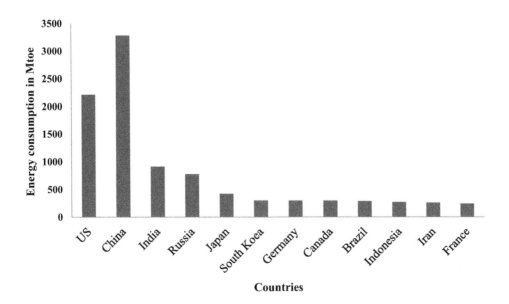

FIGURE 1.1 World Energy Consumption in Megatoe/Mtoe (one million tonne of oil equivalent).

1.8 Conclusion

Conventional energy sources such as coal and natural gas are huge contributors to climate change. Electricity production accounts for more than one-third of U.S. global warming emissions, with the majority generated by coal-fired power plants. Non-renewable electricity generation uses large amounts of scarce water and pollutes both the air and water. Reliance on conventional energy sources exposes consumers to price fluctuations and harms our health and environment. Producing electricity from coal, oil, and gas pollutes our air, soil, and water. The consequences of climate change, caused by carbon pollution from the production and burning of fossil fuels, are already visible: increased risk and duration of drought and wildfires, reduced snowpack, and extinction of vulnerable wildlife species. Renewable energy is the path to reliable, affordable, clean power.

We can power our homes and our cars with clean power, but doing so requires commitment and strategic policies to be implemented in the coming years. Renewable energy provides substantial benefits for our climate, our health, and our economy. Most renewable energy resources are carbon neutral and can be used to generate electricity, like wind, solar, and geothermal. While geothermal energy systems emit air pollutants, total air emissions are generally much lower than those of coal- and natural gas-fired power plants. Further, wind and solar energy require essentially no water to operate and thus do not pollute water resources or strain water supplies by competing with agriculture, residential use, or fish and wildlife.

In contrast, fossil fuels can have a significant impact on water resources. For example, both coal mining and natural gas drilling can pollute sources of drinking water. While these gains are significant, a more dramatic utilization of clean energy technologies will be needed if we want to address the problem of climate change. There is an urgent need to devise a renewable energy vision by identifying and extending best practices, encouraging technological innovation, analyzing costs of continued operation of existing coal-fired power plants, and promoting more effective utilization of energy efficiency and distributed renewable generation.

REFERENCES

Akhtar, A., V. Krepl and T. Ivanova. 2018. A combined overview of combustion, pyrolysis, and gasification of biomass. *Energy Fuels* 32(7): 7294–7318.

Andrea, A. 2014. Nonrenewable and renewable energy resources. https://ww2. kqed.org/quest/2014/02/13/nonrenewable-and-renewable-energy-resources–2/.

Armstrong, R. C., C. Wolfram, R. Gross, N. S. Lewis and M. V. Ramana. 2016. The frontiers of energy. *Nature Energy* 1(11): 15020.

BP plc. 2020. Natural gas – BP Statistical Review of World Energy 2020. https://www.bp.com/en/global/corporate/energy-economics/statistical-review-of-world-energy/natural-gas.html.

Center for Climate and Energy Solutions. 2020. Renewable Energy. https://www.c2es.org/.

CIA (Central Intelligence Agency). 2020. Refined Petroleum Products Consumption. https://www.cia.gov/the-world-factbook/field/refined-petroleum-products-consumption.

Consoli, C. 2019. Global CCS Institute. Bioenergy and Carbon Capture and Storage. 2019 perspective.

CT (Carbon Trust). 2006. Future Marine Energy. Results of the Marine Energy Challenge: Cost competitiveness and growth of wave and tidal stream energy. https://discomap.eea.europa.eu/.pdf.

CT (Carbon Trust). 2018. Energy management: A comprehensive guide to controlling energy use. https://www.carbontrust.com/resources/guides/energy-efficiency/energy-management.

CT (Clean Technica). 2020. Chinese Solar Perseveres During Pandemic. https://cleantechnica.com/2020/05/21/chinese-solar-perseveres-during-pandemic.

DOE (Department of Energy). 2017. History of Hydropower. https://www.energy.gov/eere/water/history-hydropower.

DOE (Department of Energy). 2020. What's the Lifespan for a Nuclear Reactor? Much Longer Than You Might Think. https://www.energy.gov.

Enerdata. 2020. Total Energy World Consumption. Global Energy Statistical Yearbook. https://yearbook.enerdata.net/total-energy/world-consumption statistics.html.

Fajardy, M., A. Köberle, N. Mac Dowell and A. Fantuzzi. 2019. BECCS deployment: a reality check. Imperial College London, Grantham Institute, Briefing Paper Number 28.

Fanchi, J.R. and C.J. Fanchi. 2016. *Energy in the 21st Century*. World Scientific Publishing Co Inc., Singapore, p. 350.

GEA (Geothermal Energy Association). 2015. The International Geothermal Market at a Glance. www.geo-energy.org/reports.aspx.

IAEA (International Atomic Energy Agency). 2021. Trend in Electricity Supplied. https://pris.iaea.org.

IEA (International Energy Agency). 2020b. Solar - Fuels & Technologies. https://www.iea.org/fuels-and-technologies/solar.

IEA (International Energy Agency). 2011. Solar Energy Perspectives: Executive Summary. http://large.stanford.edu/courses/2016/ph240/sheu1/docs/iea-solar-2011.pdf.

IEA (International Energy Agency). 2016. Ocean potential. https://www.iea.org/reports/ocean-power.

IEA (International Energy Agency). 2019. Steep decline in nuclear power would threaten energy security and climate goals. https://www.iea.org/news/steep-decline-in-nuclear-power-would-threaten-energy-security-and-climate-goals.

IEA (International Energy Agency). 2020a. Renewable energy market update: Outlook for 2020 and 2021. Fuel report.

IEA (International Energy Agency). 2020c. Global Energy Review. https://www.iea.org/reports/global-energy-review-2020.

IEA (International Energy Agency). 2020d. Renewables 2020: Analysis and forecast to 2025. Hydropower, bioenergy, CSP and geothermal. https://www.iea.org/reports/renewables-2020/hydropower-bioenergy-csp-and-geothermal.

IEA (International Energy Agency). 2020e. Solar PV – Analysis. https://www.iea.org/reports/solar-pv.

IHA (International Hydropower Association). 2016. A brief history of hydropower. https://www.hydropower.org/a-briefhistory-of-hydropower.

IRENA (International Renewable Energy Agency). 2014. Global bioenergy supply and demand projections. In D. S. A. D. G. Shunichi Nakada (Ed.), *A working paper for Remap 2030*. IRENA, Abu Dhabi, UAE.

Khan, M. A. 2007. The Geysers Geothermal Field, an Injection Success Story. Annual Forum of the Groundwater Protection Council. http://www.gwpc.org/meetings/forum/2007/proceedings/Papers/Khan,%20Ali%20Paper.pdf, retrieved 2010-01-25.

Kumar, A., G. K. Dwivedi, S. Tewari, J. Paul, V. K. Sah, H. Singh, P. Kumar, N. Kumar and R. Kaushal. 2020. Soil organic carbon pools under Terminalia chebula Retz. based agroforestry system in Himalayan foothills, India, *Current Science* 118(7): 1098–1103.

Kumar, N., N. Kumar, A. Shukla, S.C. Shankhdhar and D. Shankhdhar. 2015. Impact of terminal heat stress on pollen viability and yield attributes of rice (Oryza sativa L.). *Cereal Research Communications* 43(-4): 616–626.

Kumar, N., S.C. Shankhdhar and D. Shankhdhar 2016. Impact of elevated temperature on antioxidant activity and membrane stability in different genotypes of rice (Oryza sativa L.). *Indian Journal of Plant Physiology* 21(1): 37–43.

Liu, T., B.G. McConkey, Z. Ma, Z. Liu, X. Li and L. Cheng. 2011. Strengths, weaknesses, opportunities and threats analysis of bioenergy production on marginal land. *Energy Procedia* 5: 2378–2386.

Markandya, A. and P. Wilkinson. 2007. Electricity generation and health. *Lancet* 370 (9591): 979–990.

Martinás, K. 2005. Energy in physics and in economy. *Interdisciplinary Description of Complex Systems* 3(2): 44–58.

Moomaw, W., P. Burgherr, G. Heath, M. Lenzen, J. Nyboer and A. Verbruggen. 2011: Annex II: Methodology. In *IPCC: Special Report on Renewable Energy Sources and Climate Change Mitigation*. https://archive.ipcc.ch/pdf/special-reports/srren/SRREN_FD_SPM_final.pdf.

NASEM (National Academies of Sciences, Engineering, and Medicine). 2019. *Negative Emissions Technologies and Reliable Sequestration: A Research Agenda*. The National Academies Press, Washington, DC. https://doi.org/10.17226/25259.

Obersteiner, M. 2001. Managing climate risk. *Science* 294(5543): 786–787.

Philibert, C. 2005. The Present and Future use of Solar Thermal Energy as a Primary Source of Energy. International Energy Agency (IEA), Paris, France. http://philibert.cedric.free.fr/Downloads/solarthermal.pdf.

REN21. 2020. Renewables 2020 Global Status Report. https://www.ren21.net/wp-content/uploads/2019/05/gsr _2020_full_report_en.pdf.

Rhodes, J. S. and D. W. Keith. 2008. Biomass with capture: Negative emissions within social and environmental constraints: An editorial comment. *Climatic Change* 87(3–4): 321–328.

Richter, A., 2020. The Top 10 Geothermal Countries 2019 – based on installed generation capacity (MWe). Geothermal Energy News. https://www.thinkgeoenergy.com/the-top-10-geothermal-countries-2019-based-on-installed-generation-capacity-mwe/.

Rybach, L. 2009. Geothermal sustainability. *Geo-Heat Centre Quarterly Bulletin, Klamath Falls, Oregon: Oregon Institute of Technology* 28(3): 2–7. https://www.osti.gov/servlets/purl/1209248.

Schwab, K. 2016. The fourth industrial revolution: What it means and how to respond. https://www.weforum.org/agenda/2016/01/the-fourth-industrial-revolution-what-itmeans-and-how-to-respond.

Sentryo. 2017. The 4 industrial revolutions. https://www.sentryo.net/the-4-industrial-revolutions.

Smith, P. and J. R. Porter. 2018. Bioenergy in the IPCC Assessments. *GCB- Bioenergy* 10(7): 428–431.

Sönnichsen, N. 2020. Daily demand for crude oil worldwide from 2006–2020. Statista. https://www.statista.com/statistics/271823/daily-global-crude-oil-demand-since-2006.

Statista. 2020. China: Cumulative installed solar power capacity 2019. https://www.statista.com/statistics/279504/cumulative-installed-cpacity-of-solar-power-in-china.

Tiseo, I. 2020. Global outlook on waste to energy market value 2019–2027. Energy & Environment, Waste Management, Statista.

U.S. EIA (U.S. Energy Information Administration). 2020. Natural gas explained. Use of Natural Gas. https://www.eia.gov/energyexplained/natural-gas/use-of-natural-gas.php.

U.S. EIA (U.S. Energy Information Administration). 2021. Natural gas consumption. https://www.eia.gov/naturalgas.

U.S. EPA (U.S. Environmental Protection Agency). 2014. Electricity from Natural Gas. http://www.epa.gov/cleanenergy/energy-and-you/affect/natural-gas.html.

WCA (World Coal Association). 2009. The coal resources: A comprehensive overview of coal. https://www.worldcoal.org/file_validate.php?file=coal_resource_overview_of_coal_report(03_06_2009).pdf.

WEC (World Energy Council). 2016. World Energy Resources. https://www.worldenergy.org/wp-content/uploads/2016/10/World-Energy-Resources-Full-report-2016.10.03.pdf.

WNA (World Nuclear association). 2021. World Nuclear Power Reactors & Uranium Requirements. https://www.world-nuclear.org/information-library/facts-and-figures/world nuclear-power-reactors-and-uranium-requireme.aspx.

2

Renewable Energy: Prospects and Challenges for the Current and Future Scenarios

Manendra Singh, Amit Bijlwan, Amit Kumar, and Rajat Singh

CONTENTS

2.1 Introduction

Renewable energy refers to those that are naturally replenished and collected from natural sources for energy generation. The natural sources of energy are sun, wind, hydropower, geothermal, biomass energy, and tidal waves. Natural sources of energy are supplemented and eco-friendly to non-renewable energy such as petroleum and coal. The availability of renewable sources varies with location (Chauhan and Saini, 2014). The exploitation of natural habitats has long-term implications for the potentiality and availability of ecosystems and natural resources.

Consequently, if the habitats supplied with these resources are depleted, they will adversely affect natural resources such as food, clean air, and other ecosystem services for prolonged periods. Sustainable harvest is necessary since the natural resources are finite (Barnosky et al. 2012, Kumar et al. 2015; Kumar et al. 2020). Energy plays a significant role in developing various industrial sectors, including agriculture, manufacturing, commerce, residential, etc. In recent decades, energy demand has been steadily growing and is expected to rise continuously. Progress in renewable power sectors needs strong policy support and progressive developments in modern technologies, such as advanced biofuel plants.

In the 20th century, rapid deletion of sources of fossil fuels, natural gas, and coal was observed, and energy demand is increasing day by day in the 21st century (Quaschning, 2005; Kousksou et al. 2015). Many countries and regions worldwide have already turned to renewables as their primary energy source and aimed to supply 40%–50% of total power consumption from renewables by 2030. Presently, production of green energy (renewable energy) is growing positively, but merely the production is not enough. Its distribution system or amalgamation into energy supply networks that distribute energy to consumers using energy carriers is also required. Globally, several energy systems are diverse and contrary in terms of technological, business, financial, and cultural variations. Energy supply chains and distribution systems will need to be adapted so that they can accommodate greater supplies of renewable energy. Assimilated solutions vary with locality, scale, and the current policy of energy system and interconnected organizations and guidelines. Significant development is required in a grid system of electricity, wind power, geothermal energy, solar power, etc., for collection, transmission, and outflow management (Sims et al. 2011).

Globally, hydropower is the highest contributor to electricity generation and the primary-cum-largest renewable energy source (Masebinu et al. 2016). The transport sector is one of the fastest-growing and dynamic segments of energy based on combustion engines using fossil, leading to the rapid increase in emissions of greenhouse gases (GHG) in the last decades (Quadrelli and Peterson, 2007). However, energy demand is increasing day by day from household to development sectors. This is challenging for developing countries to fulfil the future demand of energy or expand their energy sources with efficient utilization (Krajacic, 2012). Therefore, developing countries are now focusing on developing new cost-effective renewable energy technologies to utilize the sun and wind for power generation. Technological advancement is needed to use these energy sources to an optimum level to obtain higher efficiency (Ellabban et al. 2014). Biomass-based energy systems compete economically with the fossil and fuel system. Therefore, proficient conversion technology must be selected. According to Renewable Global Status Report 2019, renewables have less growth during the last few years in the cooling, heating, and transport sectors than in the power sector. The endorsement of modern renewable energy for heating and cooling for industrial applications progressed slowly (Renewables, 2019). In 2014, investment in renewable energy increased by 16% (Masebinu et al. 2016).

Often, onshore wind and solar power are less costly than any fossil fuel alternative. Also, the running costs of existing coal-fired facilities will gradually be undermined by modern solar and wind systems. However, solar and wind energy storage technologies are older known technologies contributing the non-hydropower renewable energy (Renewables, 2015). Advancing in solar, wind, and thermal power, improving energy efficiency, and safeguarding energy for all are vital to achieve sustainable development goals by 2030. Renewable energy can be more affordable by providing a tax credit for those involved in energy production, federal assistance by government to assist lower-income families, and solar community projects to divert the investment. In the year 2019, the installed electricity capacity was increasing by more than 200 gigawatts (GW) (mostly in solar photovoltaics), the most significant increase ever in renewable energy (Joshi, 2020).

Renewable power is increasingly cost-competitive and supplementary to conventional fossil fuel-fired power plants energy. The leading causes of promoting renewable energy are to (a) minimize the industrial CO_2 emissions and (b) reduce the dependency on fossil fuel in order to improve the environmental balance. This could be the protective potential of renewables to the earth.

2.2 Prospects and Present Scenario of Renewable Energy

Energy plays a key role in driving and improving the lifestyle of humans. The consumption of energy is directly proportional to the progress of mankind. Energy storage by using different methods is becoming an increasingly prominent subject of research among the scientific communities. Renewable Energy Technologies (RET) are very important in the present and future generations for sustainable power generation and environmentally friendly use.

Renewable energy projects (REP) in developing countries have proved that they contribute in poverty mitigation and provide business and employment opportunities. Renewable energy systems played a vital role in household's electricity connection in remote areas through solar power. The United Nations Decade of Sustainable Energy for all the nation from 2014 to 2024, of the United Nations General Assembly, adopted the Agenda for Sustainable Development by 2030 under the Agenda's 17 Sustainable Development Goals. The share of renewable in electricity, heating, and the transport sector is increasing year after year. In the year 2018, it was estimated that 5% of the population in Africa and 2% of the population on the Asian continent had access to electricity through off-grid solar PV systems. Renewable energy in power generation continued its strong pace in 2018. The total power amounting to 181 GW was installed in 2018, which shows an 8% increment in total renewable installed capacity. The overall global renewable power capacity was 2,378 GW by the end of 2018 (IEA, 2018).

The private sector plays a pivotal role in driving renewable energy deployment through its procurement and investment decisions. By early 2019, 175 companies had joined RE100—committing to 100% renewable electricity targets. The government policies and planned projects are important to increase the share of renewable energy into the final energy of a particular country. The threat of climate change and challenges and mitigation strategies have also improved the renewables technology and can neutralize the balance between fossil fuels and renewable energy. Cost reductions in renewables technology and green power projects lead to increase possibilities for more investment in renewables by the government and private sectors. By the end of 2018, at least 100 cities worldwide were reportedly sourcing 70% or more of their electricity from renewables (Renewables, 2019).

Different methods are previously known for energy generation using non-conventional energy sources: solar energy, wind energy, tidal energy, biomass energy, and geothermal energy (Bhattacharjee, 2005).

2.3 Bioenergy

Globally, bioenergy contributes 9% of the total primary energy supply. The traditional use of biofuel or biomass energy in developing countries was for cooking and heating in households. In 2015, approximately 13 EJ of bioenergy was used to provide heat, representing around 6% of global heat consumption. Out of 87.5% of non-biomass energy, traditional biomass energy and modern biomass energy contribute 7.4% and 5.0%. In recent years, biomass energy has been growing fastest for electricity and transport mainly because of higher levels of policy support (IEA, 2018). The potential of bioenergy is to minimize soil, water, and air pollution through the planting of different biofuel producing herb, shrub, and tree species. In India, 26,000 saplings of *Jatropha curcus* and 500 saplings of *Pongania pinnata* have been planted under the biofuel program during 2005–2006. European countries are the largest consumers of modern bio-heat. According to the global status report: Renewables – 2019, the largest contribution of bioenergy is in the heating and cooling sector (5%), transport sector (3%), and electricity supply (2.1%).

Worldwide biofuel production increased approximately by 7% reached 153 billion litres. The United States and Brazil are dominant countries in biofuels production, contributing nearly 70% in 2018, followed by China, Germany, and Thailand, whose contribution is 3.4%, 2.9%, and 2.7%, respectively (IEA, 2018). China, India, and Japan increased their capacity by 21%, 16%, and 11%, respectively, in 2018. China provisions for further increment of bioenergy in 13th Five-Year Plan (2016–2020) of the country (WBG, 2018).

2.4 Hydropower

Hydropower is the primary source of electricity. The power is generated from the potential energy of water with the help of turbines. Hydroelectric power includes both huge hydroelectric dams and a small run of the river plants. Hydropower is highly characterized by market stability, rising industry competition, and growing demand for energy storage.

The potential and electricity generation from hydropower varies each year with shifts in weather patterns and other local conditions such as precipitation. The power is harnessed and used to drive mechanical devices used in different sectors. These pumped storage projects are being enhanced for fast response to changing grid conditions to accommodate the growing use of variable renewable power technologies. The effective integration between different renewable sources increases the potential synergies between hydropower and other renewable energy technologies (Renewables, 2019).

Hydroelectricity achieved growth in terms of capacity and instalment of new dams. In 2018, new capacity growth dams with modern technology were installed in Austria, China, and the United States. An increment of 20 GW was seen for hydropower, and total installed capacity reached around 1,132 GW. China is the leading country in hiring new hydropower capacity, representing more than 35% of new installations in 2018. The world's pumped storage capacity increased by 1.9 GW in 2018. Total storage capacity reached 160 GW, representing the massive majority of global energy storage capacity (Renewables, 2019).

2.5 Solar Energy

Solar energy is the conversion of sunlight into usable energy forms. Solar photovoltaic (PV) systems directly convert solar energy into electricity (IEA, 2018). The photovoltaic effect was first described by Edmond Becquerel in 1839, and the first viable silicon solar storage cell was developed in 1954 (Palz, 2013). PV is a modular technology and is deployed in very small quantities. Therefore, this quality allows for a wide range of applications. Cumulative solar PV capacity reached 398 GW and generated over 460 TWh, which was 2% of global power output in 2017. The solar PV system is estimated to lead renewable electricity growth by 580 GW in the next 5 years. As PV generates power from sunlight, the power output is limited when the sun is shining. However, the IEA analyzed the system amalgamation and integration of solar energy variability. Many options such as demand response, infrastructure, flexible generation, and storage exist to cost-effectively deal with solar energy challenges.

2.6 Solar Thermal Electricity

Solar energy is concentrated with the help of a focusing solar power (CSP) device. CSP concentrates the sun's rays to heat a receiver to high temperature, and then heat is transformed into electricity known as solar thermal electricity (STE). The main advantage of STE over PV is its built-in thermal storage capabilities, where CSP devices can produce electricity in continuous flow even in clouded weather. The integration of CSP solar PV and wind can improve the energy system's flexibility, and thus, CSP plants enhance energy security. Moreover, stored solar energy is converted to electricity.

2.7 Solar Heating and Cooling

Solar thermal technologies (STT) can produce heat from households scale to a very large industrial scale. Mainly STT heat is used for hot water, space heating, and industrial processes. The demand for heat is determined by the required temperature and varies with collector devices and designs. Solar energy is natural and widely available to the world. It is contributing to reduced dependence on energy imports. Solar energy diversifies the energy system against the continuously rising prices of fossil fuels. Solar technology is a one-time investment for the long term; thus, it has the potential to minimize the costs of electricity generation ranging from household level to different industrial processes. According to Technology Roadmap: Solar Heating and Cooling (SHC) report 2012, it was expected that solar energy would supply one-sixth (16.5 EJ) of the entire world's energy use for heating and cooling by the year 2050. Thus, it would save 800 MT of CO_2 emission per year. Roadmap 2012 also predicts that if intensive action is taken by governments and private agencies, annually, solar power could produce more than 16% of total energy use for low-to-moderate temperature heat and almost 17% for cooling (Renewables, 2019).

2.8 Wind Energy

Wind power is characterized by stable installations, falling prices in a competitive industry, and growing interest in offshore wind power. The Asian continent is the largest regional market, representing nearly 52% of added capacity by 2018. Falling prices help to improve markets and drive up sales. The industry is meeting challenges with ongoing technological advances, including larger turbines and improving plant efficiency and output and reducing electricity costs from wind energy. At least 12 countries produced approximately 10% of their annual electricity consumption with wind energy in 2018 (IEA, 2018; Renewables, 2019). The global transition from feed-in-terrif (FiT), which creates more competitive mechanisms, such as auctions and tenders, has resulted in intense price competition that is squeezing the entire value chain and challenging wind turbine manufacturers and developers alike. Further, wind power's success is coming with new experiments resulting from ailing designed and instigated tenders.

Electricity generation from Solar PV and wind achieved high permeation levels in 2018. Solar PV and wind are now mainstream options in the power sector. In the global scenario, maximum electricity generation is in Denmark (51%), Uruguay (36%), Ireland (29%), Germany (26%), and Portugal (24%). These countries produce more than 25% of their electricity from solar and wind. However, Spain, Greece, United Kingdom, Honduras, and Nicaragua also produced at least 12%–15% electricity from solar and wind (DEA, 2019; Renewables, 2019).

2.9 Geothermal Energy

Geothermal energy sources are unique in their location, temperature, pool depth, heating, cooling, and base load power generation from high-temperature hydrothermal sources with low to medium temperatures. Geothermal energy is also known as "direct use" in terms of energy. Geothermal power is customarily associated with diverse landform with conspicuous geothermal resources could become a viable option for many other regions and significant contributor to the power infrastructure. The physiographic and climatic variations between 60% and 90% make it a suitable technology for baseload production, generally in countries with high-temperature geothermal sources.

The new geothermal power-producing capacity came online in 2018, which is 0.5 GW, bringing global production about 13.3 GW. Turkey and Indonesia are major countries for new technology installations and contribute around two-thirds of the new capacity installed. In 2018, United States, Iceland, New Zealand, Croatia, the Philippines, and Kenya increased their geothermal energy capacity (Smith, 2017; Motyka et al., 2018).

2.10 Ocean Energy

Ocean power represents the smallest portion in the renewables market. The majority of projects are running on a relatively small scale. The resource potential of ocean energy is enormous, but this renewable sector is mainly untapped despite decades of development efforts. Net additions in 2018 were approximately 2

TABLE 2.1

Renewable Energy Capacity and Generation of Top Five Countries by 2018

	Ranking of Countries				
Renewables	**1**	**2**	**3**	**4**	**5**
Bioenergy generation	China	USA	Brazil	Germany	India
Bioenergy capacity	Iceland	USA	Brazil	India	Germany
Geothermal energy capacity	USA	Indonesia	Philippines	Turkey	New Zealand
Geothermal heat output	China	Turkey	Iceland	Japan	Hungary
Hydropower generation	China	Canada	Brazil	USA	Russia
Hydropower capacity	China	Brazil	Canada	USA	Russia
Solar PV capacity	China	USA	Japan	Germany	India
Solar PV capacity/capita	Germany	Australia	Japan	Belgium	Italy
Concentrating solar thermal power (CSP) capacity	Spain	USA	South Africa	Morocco	India
Wind power capacity	China	USA	Germany	India	Spain
Wind power capacity/capita	Denmark	Ireland	Germany	Sweden	Portugal
Total Renewable Energy	**China**	**USA**	**Brazil**	**India**	**Germany**

Source: Renewables (2019) "Global Status Report – REN21".

MW, with an estimated 532 MW of operating capacity at year's end. Financial and technological support from governments, particularly in Europe and North America, continued to reinforce private investments in ocean power technologies, especially tidal stream and wave power devices. Development activities are found worldwide but primarily concentrated in Europe, particularly in Scotland, where several arrays of tidal turbines were being deployed in 2018 (Renewables, 2019) (Table 2.1).

2.11 Renewables Growth in Asia Pacific

2.11.1 India

India is the fourth most lucrative market for renewables in the world in 2019. India will be the leading supplier for the rise in renewable energy sources in 2021, with annual increases nearly doubling by 2020. The country's target of renewable energy is ambitious, with 450 GW by 2030. This is the world's prevalent development blueprint in renewable energy. As of 30 November 2020, the overall installed aptitude for renewables is more than 90 GW with the following break up:

- Wind renewable: 38.43 GW
- Solar renewable Power: 36.91 GW
- Bio Power resources: 10.31 GW
- Small Hydro Power: 4.74 GW
- The government of India declared in 2015 a target of a total capacity of 175 GW of installed cumulative renewable power by 2022. A capacity of 85.90 GW, representing more than 23% of total installed capacity, was built until December 2019. India occupies fourth and fifth positions worldwide in the deployment of wind and solar power, respectively. The deployment of renewable energy has more than doubled between 2013 and December 2019. Over 10 million people's days of work are created in this sector every year. Solar power had risen from 2630 MW by over 14 times over the past 5 years to 37,505 MW in December 2019.
- Three solar parks having a capacity of 1,000 MW, 680 MW, and 2,000 MW are fully operational in Kurnool of Andhra Pradesh and Bhadla-II in Rajasthan and Pavagada in Karnataka.

- Twelve Biogas-based projects with a capacity of 212 kW and a corresponding capacity of 1805 m³ of biogas per day were commissioned. With this, 316 cumulative projects have been implemented in the country up to 31.12.2019 with a total electricity output of 7,166 MW and a combined overall bio-gas production of 69,500 m³ per day.

(Source: Annual Report 2019–2020, Ministry of New and Renewable Energy, Government of India)

2.11.2 China

China has been a global pioneer in renewable energy. It has enormous resources and tremendous potential for imminent growth. China has installed more new renewable energy power than Europe and the rest of the Asia-Pacific region in the year 2013 (IRENA, 2014).

- Under renewable energy map 2030, renewables in the electric/power sector will rise from 20% to nearly 40% by 2030. This means that wind and solar PV are rising rapidly, and hydroelectricity is fully deployed. This includes substantial growth of grid and transmission capacity and restructuring of the power sector.
- A guideline for energy safety released jointly by the National Development and Reform Commission (NDRC) and the National Energy Administration (NEA) in June 2020 sets out a target for 2020 of 240 GW each, for wind and solar PV, which means that wind energy will increase by 30GW, and solar energy by 36 GW in 2020; in other words, that these sources are stable and/or moderate in terms of growth (NDRC, 2020).
- For the first time, the 13th Five-year Plan set wind and solar energy generation goals, stressing the importance of incorporating renewables rather than constructing only new plants: a 420 TWh wind target and a 150 TWh solar target (NDRC, 2016).The wind is on track in 2020 to meet this goal, while solar has met its 2018 target for electricity output.
- In December 2018, the NDRC published the Action Plan for Renewable Energy Consumption (APREC) and announced its aim of "essentially solving problems in clean power integration in 2020," and set the goals of reducing wind and solar cuts below 5% in every province by that year. The proposal also calls for a "priority for unpolluted green energy use" model, stating that "renewable electricity users must be given greater obligations; grid companies and retailers should offer green electricity bundles; and users should be encouraged for sustainable energy consumption" (NDRC, 2018).
- A new energy legislation bill was issued in April 2020 by the Chinese National Energy Authority. This measure transcends the legislation on prior energy by specifically specifying that renewable power in China's energy system has a priority for development. The measure includes the implementation of a low-carbon energy grid, the phased substitution of non-fossil energy sources to replace fossil fuels, and the search for natural gas and oil. Instead of some fossil fuels, several renewables are discussed here. In the draft, it refers to clean coal production, which in China apply to newer, more efficient coal-fired plants with small particulate matter, NOx, and SO_2 emissions, including supercritical and ultra-supercritical coal-fired power but also emphasizes efficiency, environmental conservation, and ecological taxes (Energy law, 2020).

2.11.3 Japan

- During the last decade, Japan's power industry experienced significant changes, primarily due to nuclear collapse. The share in the gross electricity generation of the country dropped sharply from 25% over the fiscal year 2010 to just 6% in the fiscal year 2019, despite the major investments by the electricity companies in electricity plants. The renewable energy (RE) share almost doubled in the same time, from 10% and 19% (Shota Ichimura, 2020).
- As a result of the closure of four reactors due to different problems, nuclear energy production dropped in the first 6 months of the calendar year 2020. In the same year, RE continued to grow, reaching a remarkable 23% share in net electricity generation (Japan Atomic Industrial Forum).

- The Japan Wind Power Association (JWPA), on the other hand, has agreed for standard to abiding targets for offshore wind power installations of 10 GW by 2030 and 37 GW by 2050 (Japan Wind Power Association 2018).

2.11.4 Australia

- As a region, Australia has ratified the Paris Agreement, is targeting a 26%–28% reduction of greenhouse gas emissions by 2030, compared with 2005, and is pursuing an impressive objective of about 23% renewable energy by 2020 (sufficient projects have now been approved to meet and exceed this goal)
- In the field of solar PV in particular, it is assessed at almost 72,000 terawatt-hour (TWh) per year in eastern and south eastern Australia only. This is around 270 times Australia's total annual electricity generation (265 TWh in 2019) and 2.7 times the World's total annual electricity generation (just over 27,000 TWh in 2019) (Australian Energy Council, Solar Report, 2020).
- In June 2020, the first Renewable Energy Zones of Australia (REZ) were noted, with a 27 GW volume of new power production applications, mainly wind and solar, and storage projects connecting to the 3 GW central-west Orana REZ in New South Wales, nine times the clean energy subscription in the world. This striking result signals a strong investment interest in developing large-scale RE projects to benefit from cost savings and an efficient transmission system (Ben Willis, 2020).

2.11.5 Russia

- In the power generation sector in 2010, renewable energy use in the Russian Federation was dominated by hydropower, though bioenergy controlled heating in industry and buildings. In 2010, hydropower accounted for 70% of the total renewable energy use, and bioenergy accounted for most of the remaining 30%. In the same time period, renewable energy's share in Russia's total ultimate energy consumption was 3.6% (IRENA, 2017).
- The overall installed capacity for producing renewable energy hit 53.5 GW at the end of 2015, which accounts for around 20% of Russia's total installed capacity for generating electricity (253 GW). With 51.5 GW, hydropower accounts for virtually all this capacity, followed by bioenergy, with 1.35 GW. The installed capacity was 460 MW and 111 MW, respectively, for solar PV and onshore wind (IRENA, 2017).
- Russia-based firm Hevel Group has bespoke a 60 MW solar PV project in the Akhtuba district of the Astrakhan region. The Akhtubinskaya SES project started supplying power to the grid today, and the total estimated annual electricity generation will be 110 GWh (Tom Kenning, 2019).
- By 2024, the total state targets for an installed capacity of all RES types will be 5,863.7 MW, which includes 3,415.7 MW of wind power capacity, 2,238 MW of solar, and 210 MW of capacity from small hydroelectric power plants (Baker McKenzie, 2020).

2.12 Renewables in Transport

Transport is a dynamic sector of power consumption and contributes to the economic growth of different industrial sectors. Many countries are promoting renewables through pecuniary encouragements for electric vehicles. In 2018, Costa Rica, Germany, the Republic of Kyrgyz, and Ukraine abridged several excises for electric vehicles. The government of India allocated USD 1.3 billion (INR 87.3 billion) to inducements for battery-operated buses and other electrified vehicles such as scooters, etc. Scotland brings about USD 1.7 million (GBP 1.3 million) in allowances and credits to boost electric bike procurements. The Swedish government has come up with a new tax system for petrol and diesel vehicles and providing a tax enticement for an electric vehicle.

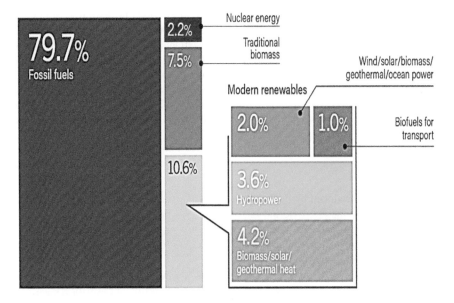

FIGURE 2.1   Renewable energy share in final energy consumption in 2017.

The Republic of Korea relieved the code of practice of manufacture and transport of hydrogen fuel cell buses. Ireland also targeted to the sale of 500,000 electric vehicles under the National Development Plan from 2018 to 2027. In 2018, the United Kingdom launched its Road to Zero strategy, with goals of ultra-low emission automobiles accounting for 50 percent to 70 percent of new car sales and 40 percent of van sales by 2030, and zero-emission cars accounting for 100 percent by 2040 (IPCC, 2014). The renewable share in final energy consumption in 2017 was approximately 21.3% (Figure 2.1).

2.13 Challenges in Renewable Energy Installation and Transmission

A powerful lever to decarbonize the power sector and reduce climate change is renewable energy sources (IPCC, 2014). In photovoltaic and wind power technologies, there has been an unprecedented increase in installed capacity, and average annual increases of 27% and 13% were recorded in the last 5 years in both the renewable sources, respectively. These variable renewables (VRE) vary from conventional generation technologies in different respects.

The new policies, programs, or technological progress indicate the barriers to renewable energy production and usage. This segment discussed some of the key hurdles and challenges in the use of renewable energy for sustainable growth. Availability of skilled human resources for various work is another problem, especially in developing countries (Mondal et al. 2010). To operate and maintain hardware equipment of renewable energy is very important to make a RE project successful, and for this purpose special training is required.

2.14 Fundamentals of Power System

Power systems are vast networks that rely on various power generators and distribute power to a wide range of loads. This means the users range from very large industrial complexes to ordinary homeowners and from a large "power transfer" to the distribution of nodes and power. These systems are thoroughly planned and prepared, and the degree of effort required to maintain such a device is greatly compounded by practically instantaneous power currents and extremely small power storage opportunities.

Solar and wind power are fundamentally distinct from conventional, nuclear, hydro, and fossil fuels. In general, these typical systems are characterized by their power capacity in terms of MW. Therefore, the sporadic problem for the system operators emerged due to length of day, season, and sometimes odd sunshine and wind create electricity fluctuation from both sources. Both are going to change unpredictably. But energy demand is not declining, so the shortage has to be addressed. Daytime or seasonal energy needs cannot be met with the available wind. Moreover, solar and wind energy need wide land expansions for usage and good sunlight and wind conditions. Substantial transport of electricity from inaccessible locations into the grid is also feasible (Stram, 2016).

2.15 Market Failures

Market failure is primarily induced by mishandling and mismanagement of the system by humans and hence described as external factors. External factors have both positive and negative effects. Very high fluctuation in unit price can lead to a negative impact on energy consumption behavioural patterns. External costs lead to sophisticated demand for harmful activities because consumers cannot bear the full charge. In the case of RE expansion, these occur as follows.

The energy-related monopoly decreases rivalry between suppliers and demand and reduces free entry and exit opportunities. Because of excessively high preliminary investment costs for RE systems, it may be too expensive or luxurious to most potential customers, typically in developing countries (Foxon, 2008). High cost in research and development in new technologies can lead to hiked prices and monopoly in the system, and growth towards achieving a sustainable future may be hindered.

2.16 Connection and Transmission

Solar and wind power generation is different from conventional methods. And generation of 1 MW power requires high expenditure in the installation process. Solar PV power systems take account of specific/unique technology to produce energy by converting direct current to alternating current. Additional steps are mandatory for certification, and solar PV installations must meet grid necessities any other circumstances. The ability to integrate electricity generated from renewable into grid supplies is governed by several factors, including (Shi et al. 2020)

- The variation with time of power generated
- Availability for a period of time
- Predictability of the energy variation
- The reliability/trustworthiness of plants
- The experience or skill of operators
- Technological advancement for integration
- Regulating the energy laws and power supply

The humongous network system must transmit the energy first from station to grid and then from grid power to end-users. Such transmission costs tend to be higher for solar and wind because of the limitation of sites for installing these power supply facilities (OECD/NEA, 2012).

2.17 Informational and Awareness Barriers

Renewable energy are site-specific, unlike fossil fuels. The output of wind turbines depends on wind speed and other associated characteristics, and only those particular favourable regions are worth the use of a wind-based RE system. Although wind regime data on the wide range may be available, comprehensive data on the local scale or region are usually unavailable. Another useful input for wind energy is topographic data, since wind turbines at high altitudes generate greater performance than comparable lower altitudinal turbines with the same hub height (Petersen et al. 1998). Local data plays an important topographical input. Therefore, the problem is the comprehensive availability of data and the enhanced modelling to provide an exact understanding of the device's performance. In the Solar power plant, accurate and reliable irradiation data is not available, without which the produced output and consequently the return on investment are difficult to measure (Sen et al. 2016).

2.17.1 Socio-Cultural Barriers

Social and cultural obstacles make issues governed by inadequate knowledge and focus on renewables. Social barriers include:

a. Willingness of farmers and other household to adopt.
b. The transition of fertile land for RE equipment/instrument installation.

Basically, farmers do not usually object to wind turbines and also see this as an additional income stream because they will continue to do agriculture and other activities under turbines. Other RE resources can, however, prevent multiple uses of land (Kotzebue et al. 2010). Dams for hydropower projects can submerge large cultivable areas and reduce the urban growth potential of land areas. Thus, social recognition and acceptance is an important factor in this (Hynes and Hanley 2006). Better understanding of RE knowledge is required with regard to social and cultural aspects of the society.

2.17.2 Policy Barriers

Different policies have been enacted and implemented to promote the use of renewables. These policies include feed-in-terrif (FiT), pricing laws, renewable portfolio standards, production incentives, and tax and trading systems (Solangi et al. 2011). Energy policy is a tactic approach in which the government adopts to resolve energy improvement issues and the upgrading and enhancement of energy engineering for its sustainable evolution, including energy production, conservation and storage, distribution, and consumption. The energy policy pertains to "the characteristics of energy policy include brief legislation, international treaties, and incentives to effective investment, strategies and guideline for energy conservation, taxation". An institutional barrier can be seen as the hegemony of existing industry and infrastructure. The energy sector has a limited number of organizations in most countries. With fewer participating organizations, the system is highly centralized, and hence it can influence policymakers for its benefits (World Bank, 2006). Many countries still have policies and regulations designed around monopoly or near-monopoly providers. These policies protect the dominant centralized energy production, transmission, and distribution, making renewable energy very difficult (Casten, 2008). A modification of existing laws and regulations is needed to introduce and promote RE technologies in the first place.

2.18 Climate Policies and Renewable Energy

Climate change and mitigation policies focused on renewable power and encourage the advancement of renewable technology and its adaptation. Effectual energy production, distribution or marketing channel, and consumption point at different levels is a principal point in global efforts to address climate change and the benefits of renewable power. Carbon emission trading systems are important policy mechanisms that can stimulate the interest in renewables to meet healthy climate goals. Policymakers focus on emission reduction from transport and industrial sectors for improving the air quality. This effort creates a strong link between renewable energy and climate policies. Intergovernmental Panel on Climate Change (IPCC) outlined several pathways such asuse of bioenergy, availability and desirability of carbon dioxide removal (CDR) technologies and carbon budget approaches that relate cumulative CO_2 emissions to global mean temperature increase and help to mitigate the threat of climate change and are reliable with a comparatively high probability of limiting the long term increase in global average temperature to 1.5°C (IPCC, 2019). Each of the pathways is characterized by reductions in the world's energy demand and increase in renewable energy efficiency through technology advancement and energy conservation. It indicates that without improving the energy efficiency, global final energy demand in 2017 would have been 12% higher than in 2000. Therefore, the average annual displacement of energy demand was below 0.7% during this period. From 2005 to 2017, the average share of renewables in total final energy consumption grew with an average annual rate of 2.9% (IEA, 2018).

2.19 Global Targets for Renewable Energy

Globally, all countries are adopting and focusing on different kinds of renewable energy targets. Denmark is the only country to target 100% for renewable energy (Renewables, 2019). However, countries such as Estonia, Finland, Latvia, and Sweden all achieved the 50% target in renewable energy for heating and cooling and transport sectors (IISD, 2018). In 2018, investment in renewables slightly decreased compared to 2017. However, the greatest advancement was seen in the proliferation of national energy efficiency action plans and targets, whereas the specific national mandates grew slowly (Table 2.2).

The European Union targeted of meeting at least 32% of its final energy consumption from renewable sources. The European Union agreement also begins a 14% share of renewable in fuels for transport energy and 1.3% annual increase in renewable heating and cooling installations under NECP. In the road transport sector, only 36% of countries are blended with biofuel energy technology. However, for the industrial sectors, 25% of renewable energy is standardized and targeted (Renewables, 2019).

TABLE 2.2

Investment on Renewable Energy and Production in 2017 and 2018

Year	2017	2018
Investment (annual) in renewable power and fuels (billion USD)	326	289
Power		
Renewable power capacity (GW)	2197	2387
Renewable power capacity (GW)	1081	1246
Hydropower capacity (GW)	1112	1131
Wind power capacity (GW)	540	591
Solar PV capacity3 (GW)	405	505
Bio-power capacity (GW)	12.1	130
Geothermal power capacity	12.8	13.3
Concentrating solar thermal power (GW)	4.9	5.5
Ocean power capacity (GW)	0.5	0.5
Bioelectricity generation (annual) TWh	532	581
Heat		
Solar hot water capacity (GW$_{th}$)	472	480
Transport		
Ethanol production (annual) billion litres	104	112
FAME biodiesel production billion litres	33	34
HVO biodiesel production (annual) billion litres	6.2	70

Source: Renewables (2019) "Global Status Report – REN21".
FAME, fatty acid methyl esters; HVO, hydro-treated vegetable oil, standard.

2.20 Conclusion

Energy demand is universally proportional to population growth and should concentrate on abundantly available renewable resources rather than conventional sources. The worldwide energy demand has been rising because of rapid growth in industrialization. Proper technologies for renewables advancement are necessary for efficient use. Such technological development can only be accomplished by coordination among green technology and market transparency along with renewable supporting policies. Thus, there are strengths, possibilities, and obstacles to the potential source of energy that can be identified in connection with this. Renewables have gained momentum in past few years, supported by modern technology, regulatory framework, and supportive policies. However, the goodwill of human beings is needed to adopt renewables in all sectors, from household to transport.

REFERENCES

Annual Report 2019–2020, Ministry of New and Renewable Energy, Government of India. https://mnre.gov. in/img/documents/uploads/file_f-1597797108502.pdf (Accessed January 15, 2021).

Australian Energy Council, Solar Report. 2020. https://www.energycouncil.com.au/reports/ (Accessed on 22/12/2020).

Baker McKenzie. 2020. New opportunities for renewable energy in Russia. https://insightplus.bakermckenzie.com/bm/energy-mining-infrastructure_1/russia-new-opportunities-for-renewable-energy-in-russia. (Accessed December 12, 2020)

Barnosky, A. D., E.A. Hadly, J. Bascompte, E.L. Berlow, J.H. Brown, M. Fortelius and N.D. Martinez. 2012. Approaching a state shift in Earth's biosphere. *Nature* 486(7401):52–58.

Bhattacharjee, C.R. 2005. Renewable energy wanted an aggressive outlook on renewable energy. *Electrical India* 45(11):147.

Casten, T.R. 2008. Recycling energy to reduce costs and mitigate climate change. In: McCracken M.C., F. Moore, J.C. Topping Jr. (eds.) *Sudden and Disruptive Climate Change: Exploring the Real Risks and How We Can Avoid Them*. London, UK: Earthscan.

Chauhan, A. and R. P. Saini. 2014. A review on Integrated Renewable Energy System based power generation for stand-alone applications: Configurations, storage options, sizing methodologies and control. *Renewable and Sustainable Energy Reviews* 38:99–120.

DEA-Danish Energy Agency. 2019. Manedlig elstatistik. Oversigtstabeller, in MonthlyElectricity Supply. https://ens.dk/en/our-services/statistics-data-key-figures-and-energy-maps/annualand- monthly-statistics (Accessed September 12, 2019)

Ellabban, O., H. Abu-Rub and F. Blaabjerg. 2014. Renewable energy resources: Current status, future prospects and their enabling technology. *Renewable and Sustainable Energy Reviews* 39:748–764.

Energy Law of the People's Republic of China [Draft for Comments], National Energy Administration, April 10, 2020. https://www.globaltimes.cn/content/1185236.shtml. (Accessed October 24, 2020)

Foxon, T. and P. Pearson. 2008. "Overcoming barriers to innovation and diffusion of cleaner technologies: some features of a sustainable innovation policy regime. *Journal of Cleaner Production* 16(1):S148–S161.

Krajacic, G. 2012. The role of energy storage in planning of 100% renewable energy system. Ph.D Thesis, Mechanical Engineering and Naval Architecture, University of Zagreb, Zagreb, Croatia.

Hynes, S. and N. Hanley. 2006. Preservation versus development on Irish rivers: White water kayaking and hydro-power in Ireland. *Land Use Policy* 23(2):170–80.

Ichimura, S. 2020, Status of Offshore Wind Power in Japan. https://www.renewable-ei.org/en/activities/column/REupdate/20200731.php. (Accessed December 08, 2020)

IEA. 2018. World Energy Balances and Statistics, 2018 edition https://www.iea.org/statistics/balances/ (Accessed July 24, 2020).

IISD - International Institute for Sustainable Development. 2018. http://www.sdg.iisd.org/news/finance-to-close-global-energy-access-gaps-to-meet-sdg-7-insufficient-seforall-report-warns/. (Accessed September 18, 2019).

IPCC, Climate Change. 2014. Synthesis report, Geneva, 2014. EIA. 2018. Installed Electricity Capacity. https://www.eia.gov/beta/international/data/browser.

IPCC-Intergovernmental Panel on Climate Change. 2019. https://www.ipcc.ch/sr15. (Accessed October 20, 2019)

IRENA. 2014. REmap 2030 Renewable Energy Prospects China- EXECUTIVE SUMMARY.

IRENA. 2017. REmap 2030 Renewable Energy Prospects for Russian Federation, Working paper, IRENA, Abu Dhabi. https://www.irena.org/remap (Accessed December 28, 2020)

Japan Atomic Industrial Forum, Current Status of Nuclear Power Plants in Japan – September 4, 2020. https://www.jaif.or.jp/cms_admin/wp-content/uploads/2020/04/jp-npps operation20200403_en.pdf (Accessed on December 2020).

Japan Wind Power Association - Toward the promotion of the introduction of offshore wind power generation (Accessed February 21, 2018).

Joshi, A. 2020. EnergyWorld-Solar photovoltaic power capacity to exceed 8,000 GW by 2050. https://energy.economictimes.indiatimes.com/news/renewable/solar-photovoltaic-power-capacity-to-exceed-8000-gw-by-2050/72302683 (Accessed December 10, 2020).

Kenning, T. 2019. Hevel commissions 60MW solar plant in Astrakhan, Russia. https://www.pv-tech.org/news/hevel-commissions-60mw-solar-plant-in-astrakhan-russia. (Accessed December 30, 2020).

Kotzebue, J., H. Bressers and C. Yousif. 2010. Spatial misfits in a multi-level renewable energy policy implementation process on the Small Island State of Malta. *Energy Policy* 38(10):5967–76.

Kousksou, T., A. Allouhi, M. Belattar, A. Jamil, et al. 2015. Renewable energy potential and national policy directions for sustainable development in Morocco. *Renewable and Sustainable Energy Reviews* 47:46–57.

Kumar, A., G. K. Dwivedi, S. Tewari, J. Paul, R. Anand, N. Kumar, P. Kumar, H. Singh and R. Kaushal. 2020. Carbon mineralization and inorganic nitrogen pools under *terminalia chebula* retz.-Based Agroforestry System in Himalayan Foothills, India. *Forest Science* 66(5):1–10.

Kumar, N., N. Kumar, A. Shukla, S.C. Shankhdhar and D. Shankhdhar. 2015. Impact of terminal heat stress on pollen viability and yield attributes of rice (Oryza sativa L.). *Cereal Research Communications* 43(-4): 616–626.

Masebinu, S., E. Akinlabi, E. Muzenda and A. Aboyade. 2016. Renewable energy: Deployment and the roles of energy storage. In *Proceedings of the World Congress on Engineering*, London, UK (Vol. 29).

Mondal, Md. A.H., L.M. Kamp and N.I. Pachova. 2010. Drivers, barriers, and strategies for implementation of renewable energy technologies in rural areas in Bangladesh—an innovative system analysis. *Energy Policy* 38:4626–34.

Motyka, M., A. Slaughter, and C. Amon. 2018. Global renewable energy trends: Solar and wind move from mainstream to preferred, https://www2.deloitte.com/us/en/insights/industry/power-and-utilities/global-renewable-energy-trends.html. (Accessed July 12, 2020)

National Development and Reform Commission (NDRC), National Energy Administration Guiding Opinions on Doing a Good Job in Energy Security in 2020. https://www.ndrc.gov.cn/xxgk/zcfb/tz/202006/t20200618_1231501.html. (Accessed December 20, 2020)

National Development and Reform Commission and National Energy Administration, December 26, 2016, http://www.nea.gov.cn/2017-01/17/c_135989417.htm. (Accessed 20, December, 2020)

National Development and Reform Commission, December 4, 2018, https://www.ndrc.gov.cn/xxgk/zcfb/ghxwj/201812/t20181204_960958.html. (Accessed November 10, 2020)

OECD, NEA. 2012. Nuclear energy and renewables: System effects in low-carbon electricity systems. Nuclear Energy Agency, 127–128.

Palz, W. 2013. *Solar Power for the World: What You Wanted to Know about Photovoltaics* (Vol. 4). Boca Raton, FL: CRC Press.

Petersen, E.L., N.G. Mortensen, L. Landberg and J.R. Hojstrup. 1998. Wind power meteorology. Part II: Siting and models. *Wind Energy* 1(2):55–72

Quadrelli, R. and S. Peterson. 2007. The energy–climate challenge: Recent trends in CO_2 emissions from fuel combustion. *Energy Policy* 35(11):5938–5952.

Quaschning, V. 2005. *Understanding Renewable Energy Systems*, 3rd ed. London: Earthscan.

Renewables. 2015. Global status report, Renewable Energy Policy Network for the 21st Century, France 978-3-9815934-6-4.

Renewables. 2019. Global Status Report – REN21, 1–336 pp. https://www.ren21.net/gsr-2019 (Accessed December 10, 2019)

REScoop. 2018. A landmark day for Europe's march towards energy democracy. https://www.rescoop.eu/news-and-events/press/a-landmark-day-for-europes-march-towards-energy-democracy (Accessed November 21, 2019)

Sen, S., S. Ganguly, A. Das, J. Sen and S. Dey. 2016. Renewable energy scenario in India: Opportunities and challenges. *Journal of African Earth Sciences* 122:25–31.

Shi R., S. Li, P. Zhang and K.Y. Lee. 2020, "Integration of renewable energy sources and electric vehicles in V2G network with adjustable robust optimization. *Renewable Energy* 153:1067–1080.

Sims, R., P. Mercado, W. Krewitt, G. Bhuyan, et al. 2011. Integration of Renewable Energy into Present and Future Energy Systems. Cambridge University Press, Cambridge, United Kingdom and New York, USA, pp. 609–706.

Smith, J.C. 2017. A major player: Renewables are now mainstream. *IEEE Power and Energy Magazine* 15(-6):16–21. (Accessed July 12, 2020).

Solangi, K. H., M.R. Islam, R. Saidur, N.A. Rahim and H. Fayaz. 2011. A review on global solar energy policy. *Renewable and Sustainable Energy Reviews* 15(4): 2149–2163.

Stram, B.N. 2016. Key challenges to expanding renewable energy. *Energy Policy* 96:728–734.

WBG. 2018. "World Bank Group announces $200 billion over five years for climate action", press release (Washington, DC). https://www.worldbank.org/en/news/press-release/2018/12/03/world-bank-group-announces-200-billion-over-five-years-for-climate-action (Accessed January 22, 2021)

Willis, B. 2020. Australia's First Renewable Energy Zone Receives 27 GW deluge of Applications -June 24, 2020. https://www.pv-tech.org/news/australias-first-renewable-energy-zone-receives-27gw-deluge-of-applications. (Accessed December 28, 2020).

World Bank. 2006. Reforming power markets in developing countries: What have we learned? Washington, DC, USA.

3

Green Technology in Relation to Sustainable Agriculture: A Methodological Approach

Ishwar Prakash Sharma and Chandra Kanta

CONTENTS

3.1 Introduction

Sustainable agriculture is indispensable in the recent world as it has the potential to enhance our daily agricultural needs without causing environmental damage, which is one of the key solutions to increase the augmented food demand. Various chemicals are being used for agricultural productivity; the continuous use of these chemical fertilizers, pesticides, and herbicides has led to lower down soil fertility, low agricultural productivity, soil damage, loss of biodiversity, unfavourable economic returns, food poisoning, and serious environmental hazards. The quantity and quality of agricultural products affected by the use of such chemicals in agriculture practices have been discussed by many of the researchers under the agriculture environment relationships (Meena et al. 2016; Dwivedi et al. 2017; Sharma and Sharma 2020). Agriculture is one of the most significant factors contributing to the economic growth of a nation, and sustainable agriculture economic development is the need of the hour. Hence, a major focus in the coming decades would be on safe and eco-friendly methods, policies, and programs to exploit sustainable agricultural production (Whittinghill and Rowe 2011). Despite many techniques such as genetic engineering, micro-flora involvement, rhizosphere engineering, transgenic plant production, etc., being used recently, recently researches focus on green technology, which could have several benefits in agriculture sectors in the coming decades. For the first time, the United Nations Asian and Pacific Centre for Agricultural Engineering and Machinery (APCAEM) has been taking initiative toward green technology to promote agro-based environment friendly technology and the development of sustainable agriculture for eradication of poverty and guaranteeing environmental sustainability. Some agroecosystem processes are not associated with the known land or field-level knowledge, so we need to understand all the agroecological-based processes to the land level that directly involved in the soil function and regulation along with other processes like pest control. In this scenario, we need to understand and increase our knowledge of green technology for developing sustainable agriculture with minimal risk to the environment. Green technology is the application of different environmental sciences and monitoring, green chemistry, and electronic devices to monitor that conserve the natural environment and resources and lead to curbing of the negative impact of human involvement (Boye and Arcand 2012). Commonly, the term green technology is an umbrella term that describes the use of science and technology to improve production for a more environmentally friendly nature. Under this field, we include technology infrastructure for waste recycling, water purifying, energy production, and natural resources conservation. As per the report of United Nations (2018), the green technology processes surpassed $200 billion in 2017, with $2.9 trillion invested in various solar and wind power sources since 2004; this report also states the China is largest global investor in this sector with approximately $126 billion invested in 2017. On the basis of all the above-mentioned techniques, we need to elaborate and develop new models for sustainable agriculture that follow the eco-friendly nature. This chapter focuses on the various eco-friendly processes that directly involve agricultural development. These eco-friendly techniques or green technologies will be achieving great importance toward sustainable agricultural development for present and future generations.

3.2 Green Technology

The green technology is generally the practical knowledge of technological applications toward the sustainable development. In this context, the technologies allow people to become more efficient and to think about various strategies that were not possible earlier. Beneficially, the technology can successfully link different programmes of a country that can solve socio-economic problems. The agricultural productivity increase is possible through various environmental friendly and beneficial technologies that provide a number of benefits such as reduced water and soil pollution, carbon sequestration, increase in biodiversity and lifespan, and energy saving (Whittinghill and Rowe 2011). Green technology is a good link between poverty reduction and productivity growth. In this context, the local knowledge from any area can be merge with technology for better production; most of the traditional knowledge is limited in a particular area that is much better for yield and production; hence, we need to combine this knowledge to the technology that reduces poverty and improves production (Sharma et al. 2020). This combined knowledge or ecological processes directly involve the food production along with nutrient recycling, microclimate regulation, suppression of undesirable entities, detoxification of noxious chemicals, etc. (Whittinghill and Rowe 2011). Most of the traditional knowledge by the local farmers offer a wide array of management options that enhance functional biodiversity in the field. Most of the agriculture based on chemical fertilizers throughout the world has now been minimized by the efforts of green technology. The technologies used in green technology should be in harmony with natural and ecological principles. There are some technologies associated with harmful products (petroleum products) that might be reduced by the green technology for sustainable development. Green technology covers a broad group of methods and materials for generating energy to non-toxic cleaning products (Pyakuryal 2012).

3.3 Green Technology Association

The respective governments from various countries are involved in several environmentally-sensitive projects because the environmental impact assessment must be assessed for sustainable development. In Asia, the wastewater and sewage disposal are the major threats to human health; along with these, domestic waste, commercial properties, and industry or agriculture generate potential contaminants that are major threats to the world environment that need to minimised or recycled, which can be possible by the green technologies. Generally, the green technology is affordable for renewable energy, such as sunlight, wind, water, geothermal heat, etc., that is useful in solar power, wind power, hydroelectricity/micro hydro, biomass, and biofuels. The non-renewable energy is also associated with the green technology. The understandings of these energies are very useful for sustainable development that includes environmental, economic, and socio-political sustainability (Pyakuryal 2012). The various association of green technology with agriculture have been mentioned in the next section.

3.4 Different Agriculturally Important Strategies for Green Technology Development

3.4.1 Organic Agriculture in Relation to Green Technology

Organic agriculture is a very old concept, but it has been hotly debated since the last two decades as a potential approach to ensure food security (Pickett 2013). The organically managed farmlands or farms are gradually increasing food secularity in the global market. There are many safe and secures strategies that hinder the exclusive use of synthetic chemicals in the farm land and use of genetically modified organisms that ultimately improve soil health, ecosystem and crop productivity (Saffeullah et al. 2021). The organic agriculture is based on management practices, ecological harmony enrichment, and sustainability for attaining the objectives of environment, economy, and society (Pamela 2010). Recently, a total of 172 out of 227 countries are involved in organic agriculture, contributing 76% of the total world's production. In this series, Europe contributes 100% to organic agriculture followed by Asia (79%) and Africa (70%). Currently 0.99% of cultivated land is organically used (Saffeullah et al. 2021). Chadha and Srivastava (2020) mentioned in their study that 103 countries out of 186 contributed in organic agriculture, while 1.5% of agricultural land is organic. In the report of International Federation of Organic Agriculture Movements (IFOAM) (2020), India was the largest organic producer in the world in 2018 followed by Uganda and Mexico. In this scenario, organic agriculture helps manage soil fertility along with the land's essential features that ultimately improve plant yield and productivity (Koutroubas and Damalas 2016). Along with agricultural sustainability, organic agriculture faming for a sustainable environment for green technology evidences many results. The problem related to chemicals that have been frequently used in soil are that they directly contribute to the pollution of soil and water and these issues have been resolve by using organic agriculture in the several studies (Ngatia et al. 2019). In the series of green technology, several reports authenticate the importance of organic agriculture for global warming and climate change (IPCC 2007a, b; FAO 2008, 2011).

3.4.2 Role of Cyanobacteria for Sustainable Green Technology

Cyanobacteria also known as blue green algae are the most abundant group of organisms on the earth and are autotrophic and mostly found in marine, rarely in, freshwater. Marine water is the richest source for their cultivation (Hoekman et al. 2012). They are microscopic, and unicellular and grow in colonies. About 150 genera of cyanobacteria have been identified till date. They have natural potential to fix atmospheric nitrogen through their heterocysts; hence, they could be used as a biofertilizer for cultivation of economically important crops. This ultimately enhances production and soil fertility (Zahra et al. 2020). Cyanobacteria are emerging microorganism for sustainable agricultural development that control nitrogen deficiency, improve soil aeration, and enhance water holding capacity (Zahra et al. 2020). Common usable Cyanobacteria *Nostoclinkia, Anabaena variabilis, Aulosira fertilisima, Calothrix* sp., *Tolypothrix* sp., and *Scytonema* sp. are present in the rice crop cultivation area. *Nostoc* and *Anabaena* that are present in surface of soil and rocks can fix up to 20–25 kg/ha atmospheric nitrogen. There are various symbiotic associations of Cyanobacteria with other group of organisms such as *Azolla-Anabaena* association in the paddy field. The applications of these biofertilizers have been reported in maize, tomato, sugarcane, oats, barley, radish, cotton, lettuce, and chilli (Thajuddin and Subramanian 2005). Cyanobacteria play a major role in soil build-up, maintenance, and fertility. Phosphorus is the second most important nutrient in agriculture that can increase due to the presence of Cyanobacteria owing to improvement in organic acid production. Globally, about one billion ha of agriculture land is affected by salinization that could be reclaimed by various biological, physical, and chemical remediation processes; in this regards Cyanobacteria are the good biological alternative for soil salinization treatment due to their atmospheric nitrogen fixing abilities that produce an extracellular matrix and compatible solutes by which they help in reduction and decay of weeds and increase soil biomass (Li et al. 2019). Cyanobacteria have the capacity to tolerate ultraviolet radiation, desiccation, high or low temperatures; hence, it is an advantage for competitors/neighbours in their natural habitats. Overall, the Cyanobacteria are an important environmentally, agriculturally, and economically sustainable option. The reported species of Cyanobacteria used as biofertilizer in various agriculture crops mentioned in Table 3.1. Most of the research on Cyanobacteria as biofertilizers and plant-promoting activities are studied in Asian and some parts of African continent because they have the largest cereals-producing countries. Cyanobacteria have potential to change soil conditions, which is most important reason to promote these types of studies (Múnera-Porras et al. 2020).

TABLE 3.1

Cyanobacteria and Their Symbiotic Agricultural Crops

Cyanobacteria	Crop	References
Calothrix elenkinii, Anabaena spiroides, A. variabilis, A. torulosa, A. osillarioides	*Oryza sativa*	Khatun et al. (2012), Prasanna et al. (2013a), Priya et al. (2015), Kumar et al. (2015), Wang et al. (2015), Shahane et al. (2019), Kumar et al. (2019)
Calothrix sp., *Anabaena laxa, A. torulosa, A. doliolum, Nostoc carneum, N. piscinale*	*Glycine max*	Prasanna et al. (2014)
Calothrix sp., *Anabaena laxa, A. torulosa, A. doliolum, Nostoc carneum, N. piscinale*	*Vigna radiata*	Dey et al. (2017a), Prasanna et al. (2014)
Anabaena torulosa, A. laxa	*Cicer arietinum*	Bidyarani et al. (2016)
Anabaena laxa, A. azollae, A. oscillarioides, A. torulosa, Calothrix crustacean	*Triticum aestivum*	Rana et al. (2012, 2015), Nain et al. (2010), Manjunath et al. (2011), Swarnalakshmi et al. (2013), Prasanna et al. (2013b)
Calothrix elenkinii	*Vigna unguiculata*	Dey et al. (2017b)
Spirulina platensis, Nostoc carneum, N. piscinale, N. muscorum, N. calcicola, Anabaena torulosa, A. doliolum, A. oryzae, Oscillatoria nigra, O. princeps, O. curviceps, Schizothrix vaginata, Lyngbya gracilis, Phormidium dimorphum, Calothrix clavata, Aulosora prolifica, Stigonema dendroideum, S. varium, Gloeocapsa calcarea, Tolypothrix tenuis	*Zea mays*	Dineshkumar et al. (2019), Prasanna et al. (2016a), Anand et al. (2015)
Anabaena sp., *Calothrix* sp.	*Abelmoschus esculentus*	Manjunath et al. (2016)
Spirulina platensis, Nostoc muscorum, A. oryzae	*Apium graveolens*	Rashad et al. (2019)
Spirulina platensis	*Raphanus sativus*	Godlewska et al. (2019)
Nostoc entophytum, Oscillatoria angustissima, Anabaena torulosa	*Pisum sativum*	Osman et al. (2010), Verma et al. (2016)
Calothrix elenkinii, Anabaena laxa, A. torulosa	*Gossypium hirsutum*	Prasanna et al. (2016b)
Anabaena vaginicola, Nostoc calcicola	*Cucurbita maxima*	Shariatmadari et al. (2013)
Anabaena vaginicola, Nostoc calcicola	*Cucumis sativus*	Shariatmadari et al. (2013)
Anabaena vaginicola, Nostoc calcicola	*Solanum lycopersicum*	Shariatmadari et al. (2013)

3.4.3 Role of Plant Growth-Promoting Rhizobacteria (PGPRs) in Sustainable Green Technology

Among all the sectors, agriculture releases the maximum amount of chemical pollutants because of the multiple uses of synthetic pesticides and chemical fertilizers that cause great destruction in the ecosystem and lead to health problems. The excessive uses of these chemicals cause environmental pollution due to bioaccumulation and biomagnification processes. The heavy metals like mercury, cadmium, and lead are retained in plants via soil and pass to other organisms through the food chain. Diseases like stomach cancer, vector-borne diseases, and methemoglobinemia have been reported due to this above process in infants and adults (Katiyar et al. 2016). In case of fertilizers, there are various N_2-based fertilizers that produce Nitrous oxide (N_2O), a major greenhouse gas leads to global warming (Rai et al. 2020). Hence, to manage all these problems, PGPRs are alternatives. PGPRs enhance the soil fertility for the availability of nutrient in this scenario. Sharma et al. (2017) gave term 'Heart of Soil' to the PGPRs. PGPRs diversity varies on the soil type and conditions, plant type, and nutrient availability in rhizosphere (Sharma et al. 2018; Tian et al. 2020). They act directly and indirectly against the biotic as well as abiotic stresses (Pathania et al. 2020; Kour et al. 2020). The mechanism of PGPRs elaborated in Figure 3.1. They are another type of biofertilizer that directly involve plant growth and lead to enhancement for sustainable agriculture and green technology. In the series of biofertilizer, they increase nitrogen fixation, phosphate solubilization, sederophore production, and their cellulolytic activities lead to plant promotion, productivity, and yield increase due to easier uptake of soil nutrients that are directly involve in maintaining the soil quality in the rhizosphere zone (Riaz et al. 2020). For sustainable agriculture, they improve plant growth in terms of seed germination, shoot and root development, increased biomass, and reduced disease incidence, which is an outcome of biofertilizer application (Pathania et al. 2020). Biopesticide is an alternative eco-friendly application of PGPRs for management of plant weeds and various diseases that involve in agriculture loss. In this series, PGPRs like *Bacillus, Pseudomonas, Serratia, Arthrobacter,* and *Stenotrophomonas* are involved in the production of various volatile compounds and other metabolites as enzymes, proteins, antibiotics, etc. also that enhance agriculture growth and improve soil health (Vejan et al. 2016). Overall, the PGPRs are good mate of agriculture and have thepotential in enhancement of agriculture production.

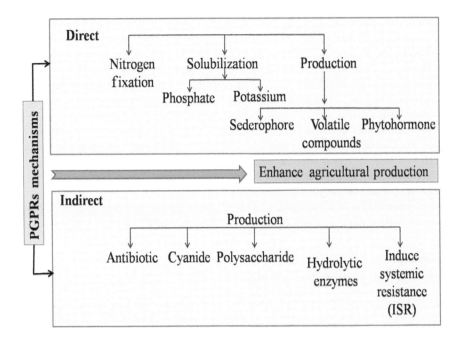

FIGURE 3.1   A diagrammatic representation of PGPRs mechanisms.

3.5 Below-Ground Alteration for Sustainable Agriculture

The studies on below-ground biota have made great efforts in this decade. Most of the studies have investigated soil microbiota focusing on key issues such as structure, composition, and function of below-ground diversity along with their environmental factors like changing environment, climatic conditions, soil pollution, soil disturbance, host plant functions, etc. (Yuan et al. 2020). The below-ground biota vary in agricultural and forest system because in agricultural system the soil is properly rotated via tillage along with various agricultural inputs like fertilizers application, pesticides, irrigations, etc. that build up the consortia of soil microbiota, while in forest system, these do not happen; in this scenario, the energy flow in forest systems is through photosynthesis and nutrient mobilization via plant to soil helps to maintaining the soil biota consortia. In such undisturbed natural processes, a powerful metagenomic approach was illustrated by Colagiero et al. (2017). Structural and functional modifications in soil biota have potential impact on the above-ground ecosystems. Recently, new gene editing and synthetic biology tools offer another task to the rhizosphere engineers to improve rhizosphere health along with these breeding or engineering plant traits gain potential support for rhizosphere. The root exudates support the manipulation of below-ground microbial diversity; many genes controlling exudates have been identified that need to be altered or redesigned for better expression toward rhizosphere improvement, which can directly affect the above-ground productivity. Biochemically, the extracellular enzymes are higher in rhizosphere as compare with root-free soil, similar to microbial biomass and activity due to rhizosphere effect. The most recent strategy to alter the rhizosphere is the multi-omic approach that is a modern and holistic approach to rhizosphere biology that elaborates the different datasets as genome, proteome, transcriptome, metabolome, etc. (Köberl et al. 2020). These abovementioned approaches could alter the below-ground biota and resolve various rhizosphere problems that ultimately lead to an eco-friendly and sustainable agriculture.

3.6 Role of Fungal Microbiota in Relation to Green Technology

Fungi are more difficult to observe and control, and avoiding their occurrence is impossible because they could be hardly identified due to their low concentrations (Cornejo et al. 2015). Among all the fungi, some are harmful or pathogenic to the agriculture crops that should be manage by eco-friendly methods. In this scenario, many biotic organisms including bacteria, fungi, microalgae, compost, etc., have been studied from time to time by the rhizosphere biologist. In most of the studies various fungi have been reported suitable for managing it. Some agriculturally important fungi with their role for sustainable agriculture are mentioned in Table 3.2. The mycorrhizal fungi are gaining importance as they can manage disease (Selvaraj and Thangavel 2021) while nonmycorrhizal fungi such as *Trichoderma* spp. (Druzhinina et al. 2011), *Piriformospora indica* (Waller et al. 2005), *Penicillium simplicissimum* (Alam et al. 2011), *Saccharomyces* spp. (Martino et al. 2003), *Aspergillus* spp. (Alves et al. 2010), *Pichia* spp. (Fiori et al. 2008), *Candida oleophila* (Porat et al. 2003), *Mortierella* spp. (Ozimek and Hanaka 2020), etc. have been well studied. Along with the disease control, they promote plant health and crop productivity by reducing plant stresses (Kour et al. 2020), thus helping plants survive or tolerate global climatic change. Hence, for green technological approach, the fungi should be a potential agriculture-based strategy to manage agriculture in relation to disease control, stress reduction, soil management, crop production, and most recently acclimatization towards global climate change.

3.7 Conclusion and Future Perspective

The development and growth of synthetic chemicals have significantly enhanced the agriculture sector, but their use causes environmental threats, which is the main disadvantage of these chemicals. In this scenario, we need to reduce their utilization for sustainable agriculture towards future sustainable perspective. Hence, the various approaches and policies regarding green technologies need to be applied

TABLE 3.2

Fungal Microbiota and Their Role for Sustainable Agriculture

Fungi	Role	References
***Ectomycorrhizal* fungi** (*Pisolithus tinctorius, Amanita* sp., *Tuber* sp., *Lactarius* sp., *Elaphomyces* sp., *Pisolithus* sp., *Piriformospora indica, Cenococcum* sp., *Rhiizopogon* sp.)	Biofertilizer	Alves et al. (2010)
Arbuscular mycorrhiza (*Glomus viscosum, G. aggregatum, G. cerebriforme, G. deserticola, G. globiferum, G. halonatum, G. intraradices, G. manihotis, G. microcarpum, G. monosporum, G. mosseae, G. radiatum, G. versiforme*)	Biofertilizer	Adholeya et al. (2005), da Silva (2006), Rai et al. (2013), Anderson and Cairney (2007), Pal et al. (2015)
***Aspergillus* spp.** (*A. awamori, A. fumigatus, A. melleus, A. niger, A. terreus, A. tubingensis*)	Biofertilizer	Manoharachary et al. (2005)
***Penicillium* spp.** (*P. adametzioides, P. albidum, P. bilaji,, P. citrinum, P. expansum, P. frequentans, P. italicum, P. oxalicum, P. roqueforti, P. rubrum, P. simplicissimum, P. viridicatum*)	Biofertilizer, Biocontrol	Khokhar et al. (2013), Alam et al. (2011), Meesala and Subramaniam (2016)
***Trichoderma* spp.** (*T. asperellum, T. harzianum T. koningii, T. pseudokoningii, T. virens, T. viride*)	Biofertilizer, Biocontrol	Manoharachary et al. (2005), Rahman et al. (2012)
***Fusarium* spp.** (*F. moniliforme, F. oxysporum, F. solani, F. udam*)	Biofertilizer	Harman et al. (2004)
***Mucor* spp.** (*M. hiemalis, M. mucedo, M. ramosissimus*)	Biofertilizer	Manoharachary et al. (2005)
***Candida* spp.** (*C. krissii, C. oleophila, C. scotti*)	Biofertilizer, Biocontrol	Porat et al. (2003)
***Tritirachium* spp.** (*T. album, T. egenum*)	Biofertilizer	Pal et al. (2015)
***Oidiodendron* spp.** (*O. maius*)	Biofertilizer	Martino et al. (2003)
***Pleurotus* spp.** (*P. flabellatus, P. ostreatus*)	Biofertilizer	Whitelaw (1999)
***Chaetomium* spp.** (*C. bostrychodes, C. olivaceum*)	Biofertilizer	Pal et al. (2015)
***Pichia* spp.** (*P. angusta, P. membranifaciens*)	Biocontrol	Fiori et al. (2008)
***Lentinus* spp.** (*L. conatus*)	Biocontrol	Lakshmanan et al. (2008)
***Streptomyces* spp.** (*S. lydicus*)	Biocontrol	Yuan and Crawford (1995)
***Leucosporidium* spp.** (*L. scottii*)	Biocontrol	Vero et al. (2013)
***Saccharomycopsis* spp.** (*S. schoenii*)	Biocontrol	Hashem et al. (2012)

for sustainable agriculture that promotes eco-friendly agriculture. Green technology, such as use of beneficial microbiota and their products, may be significant alternatives solutions to these problems. Among various microbiota the cyanobacteria, PGPRSs, and fungi along with their formulations have great potential to deal with the various agricultural problems as stresses, nutrient deficiency, disease tolerance, etc. In many of the previous studies, the beneficial roles of these organisms have been well documented. Along with these, the traditional agricultural practices have great potential toward sustainable agriculture. Further continuous research needs to be done to identify and isolate the beneficial microbiota for their multifarious characteristics that should be of further use in agriculture for sustainable agriculture and environmental practices. All the mentioned practices improve plant growth and productivity and lead to an eco-friendly environment.

REFERENCES

Adholeya, A., P. Tiwari, and R. Singh. 2005. Large-scale inoculum production of arbuscular mycorrhizal fungi on root organs and inoculation strategies. In *Vitro Culture of Mycorrhizas, Soil Biology*, edited by Declerck, S., J. A. Fortin, D. G. Strullu, pp. 315–338. Springer, Berlin, Heidelberg.

Alam, S. S., K. Sakamoto, and K. Inubushi, 2011. Biocontrol efficiency of Fusarium wilt diseases by a root-colonizing fungus Penicillium sp. *Soil Science and Plant Nutrition* 57: 204–212.

Alves, L., V. L. Oliveira, and G. N. S. Filho. 2010. Utilization of rocks and ectomycorrhizal fungi to promote growth of eucalypt. *Brazilian Journal of Microbiology* 41: 676–684.

Anand, M., K. Baidyanath, and N. Dina. 2015. Cyanobacterial consortium in the improvement of maize crop. *International Journal of Current Microbiology and Applied Sciences* 4(3): 264–274.

Anderson, I. C., and J. W. G. Cairney. 2007. Ectomycorrhizal fungi: exploring the mycelial frontier. *FEMS Microbiology Reviews* 31: 388–406.

Bidyarani, N., R. Prasanna, S. Babu, et al. 2016. Enhancement of plant growth and yields in Chickpea (Cicer arietinum L.) through novel cyanobacterial and biofilmed inoculants. *Microbiological Research* 188–189: 97–105.

Boye, J., and Y. Arcand. 2012. *Green Technologies in Food Production and Processing*. Springer Science & Business Media, Berlin.

Chadha, D., and S. K. Srivastava. 2020. Growth performance of organic agriculture in India. *Current Journal of Applied Science and Technology* 39: 86–94.

Colagiero, M., I. Pentimone, L. Rosso, et al. 2017. A metagenomic study on the effect of aboveground plant cover on soil bacterial diversity. In *Soil Biological Communities and Ecosystem Resilience, Sustainability in Plant and Crop Protection Series*, edited by Lukac, M., P. Grenni, and M. Gamboni, pp. 97–106. Springer International Publishing AG, Switzerland.

Cornejo, H. A. C., L. M. Rodriguez, A. G. Vergara, et al. 2015. Trichoderma modulates stomatal aperture and leaf transpiration through an abscisic acid-dependent mechanism in Arabidopsis. *Journal of Plant Growth Regulation* 34 (2): 425–434.

da Silva, E. 2006. *Handbook of Microbial Biofertilizers*. CRC Press, Boca Raton, FL.

Dey, S. K., B. Chakrabarti, R. Prasanna, et al. 2017a. Elevated carbon dioxide level along with phosphorus application and cyanobacterial inoculation enhances nitrogen fixation and uptake in cowpea crop. *Archives of Agronomy and Soil Science* 63 (13): 1927–1937.

Dey, S. K., B. Chakrabarti, R. Prasanna, et al. 2017b. Productivity of mungbean (Vigna radiata) with elevated carbon dioxide at various phosphorus levels and cyanobacteria inoculation. *Legume Research* 40: 497–505.

Dineshkumar, R., J. Subramanian, J. Gopalsamy, et al. 2019. The impact of using microalgae as biofertilizer in maize (Zea mays L.). *Waste and Biomass Valorization* 10: 1101–1110.

Druzhinina, I. S., V. Seidl-Seiboth, A. Herrera-Estrella, et al. 2011. Trichoderma: The genomics of opportunistic success. *Nature Reviews Microbiology* 9: 749–759.

Dwivedi, N., P. Kumar, S. K. Dwivedi, et al. 2017. Green technology: An eco-friendly approach towards sustainable agriculture. *G-Journal of Environmental Science and Technology* 4 (4): 31–33.

FAO (Food and Agriculture Organization). 2008. *Organic Agriculture and Climate Change*. FAO, Rome.

FAO (Food and Agriculture Organization). 2011. Organic Agriculture Retrieved 17 Nov, 2011.

FiBL and IFOAM Organics International. 2020. The World of Organic Agriculture: Statistics and Emerging Trends. Research Institute of Organic Agriculture (FiBL), Frick, Switzerland.

Fiori, S., A. Fadda, S. Giobbe, et al. 2008. Pichia angusta is an effective biocontrol yeast against postharvest decay of apple fruit caused by Botrytis cinerea and Monilia fructicola. *FEMS Yeast Research* 8: 961–963.

Godlewska, K., I. Michalak, P. Pacyga, et al. 2019. Potential applications of cyanobacteria: Spirulina platensis filtrates and homogenates in agriculture. *World Journal of Microbiology and Biotechnology* 35(6): 80.

Harman, G. E., C. R. Howell, A. Viterbo, et al. 2004. Trichoderma species- opportunistic, avirulent plant symbionts. *Nature Reviews Microbiology* 2: 43–56.

Hashem, M., R. Agami, and S. A. Alamri. 2012. Effect of soil amendment with yeasts as bio-fertilizers on the growth and productivity of sugar beet. *African Journal of Agricultural Research* 7: 6613–6623.

Hoekman, S. K., A. Broch, C. Robbins, et al. 2012. Review of biodiesel composition, properties, and specifications. *Renewable and Sustainable Energy Reviews* 16(1): 143–169.

IPCC. 2007a. Third assessment report mitigation. IPCC, Switzerland.

IPCC. 2007b. Contribution of working group II to the fourth assessment report of the intergovernmental panel on climate change – summary for policymakers. Retrieved on Nov 10, 2008.

Katiyar, D., A. Hemantaranjan, and B. Singh. 2016. Plant growth promoting rhizobacteria-an efficient tool for agriculture promotion. *Advances in Plants & Agriculture Research* 4(6): 00163.

Khatun, W., M. M. Ud-Deen, and G. Kabir. 2012. Effect of cyanobacteria on growth and yield of boro rice under different levels of urea. *Rajshahi University Journal of Life & Earth and Agricultural Sciences* 40: 23–29.

Khokhar, I., M. S. Haider, I. Mukhtar, et al. 2013. Biological control of Aspergillus niger, the cause of Black-rot disease of Allium cepa L. (onion), by Penicillium species. *Journal of Agrobiology* 29: 23–28.

Köberl, M., P. Wagner, H. Müller, et al. 2020. Unraveling the complexity of soil microbiomes in a large-scale study subjected to different agricultural management in Styria. *Frontiers in Microbiology* 11: 1052.

Kour, D., K. L. Rana, T. Kaur, et al. 2020. Microbe-mediated alleviation of drought stress and acquisition of phosphorus in great millet (Sorghum bicolor L.) by drought-adaptive and phosphorus-solubilizing microbes. *Biocatalysis and Agricultural Biotechnology* 23: 101501.

Koutroubas, S. D., and C. A. Damalas. 2016. Morpho-physiological responses of sunflower to foliar applications of chlormequat chloride (CCC). *Bioscience Journal* 32(6): 1493.

Kumar N, N. Jeena, and H. Singh 2019. Elevated temperature modulates rice pollen structure: A study from foothill Himalayan Agro-ecosystem in India. *3Biotech* 9: 175.

Kumar, N., N. Kumar, A. Shukla, S. C. Shankhdhar and D. Shankhdhar. 2015. Impact of terminal heat stress on pollen viability and yield attributes of rice (Oryza sativa L.). *Cereal Research Communications* 43(-4): 616–626.

Lakshmanan, P., R. Jagadeesan, A. Sudha, et al. 2008. Potentiality of a new mushroom fungus Lentinus connatus Berk for the production of biomanure from sugarcane trash (Saccharum officinarum L.) and its impact on the management of groundnut root rot diseases. *Archives of Phytopathology and Plant Protection* 41: 273–289.

Li, N., X. Zhang, C. Zhang, et al. 2019. Real-time crop recognition in transplanted fields with prominent weed growth: A visual-attention-based approach. *IEEE Access* 7: 185310–185321.

Manjunath M., R. Prasanna, P. Sharma, et al. 2011. Developing PGPR, consortia using novel genera Providencia and Alcaligenes along with cyanobacteria for wheat. *Archives of Agronomy and Soil Science* 57 (8): 873–887.

Manjunath, M., A. Kanchan, K. Ranjan, et al. 2016. Beneficial cyanobacteria and eubacteria synergistically enhance bioavailability of soil nutrients and yield of okra. *Heliyon* 2: e00066.

Manoharachary, C., K. Sridhar, R. Singh, et al. 2005. Fungal biodiversity: Distribution, conservation and prospecting of fungi from India. *Current Science* 89: 58–71.

Martino, E., S. Perotto, R. Parsons, et al. 2003. Solubilization of insoluble inorganic zinc compounds by ericoid mycorrhizal fungi derived from heavy metal polluted sites. *Soil Biology and Biochemistry* 35: 133–141.

Meena, M. K., S. Gupta, and S. Datta. 2016. Antifungal potential of PGPR, their growth promoting activity on seed germination and seedling growth of winter wheat and genetic variabilities among bacterial isolates. *International Journal of Current Microbiology and Applied Science* 5(1): 235–243.

Meesala, S., and G. Subramaniam. 2016. Penicillium citrinum VFI-51 as biocontrol agent to control charcoal rot of sorghum (Sorghum bicolor (L.) Moench). *African Journal of Microbiology Research* 10: 669–674.

Múnera-Porras, L. M., S. Garcia-Londono, and L. A. Rios-Osorio. 2020. Action mechanisms of plant growth promoting cyanobacteria in crops in situ: A systematic review of literature. *International Journal of Agronomy* 2020: 1–9.

Nain, L., A. Rana, M. Joshi, et al. 2010. Evaluation of synergistic effects of bacterial and cyanobacterial strains as biofertilizers for wheat. *Plant and Soil* 331: 217–230.

Ngatia, L., J. M. Grace, D. Moriasi, et al. 2019. Nitrogen and phosphorus eutrophication in marine ecosystems. Monitoring of Marine Pollution, pp.1–17.

Osman, M. E. H., M. M. El-Sheekh, A. H. El-Naggar, et al. 2010. Effect of two species of cyanobacteria as biofertilizers on some metabolic activities, growth, and yield of pea plant. *Biology and Fertility of Soils* 46: 861–875.

Ozimek, E., and A. Hanaka. 2020. Mortierella species as the plant growth-promoting fungi present in the agricultural soils. *Agriculture* 11(1): 1–1.

Pal, S., H. B. Singh, A. Farooqui, et al. 2015. Fungal biofertilizers in Indian agriculture: perception, demand and promotion. *Journal of Eco-Friendly Agriculture* 10: 101–113.

Pamela, C. 2010. Guide for organic crop producers. National Center for appropriate technology (NCAT) agriculture specialist p. 64.

Pathania, P., A. Rajta, P. C. Singh, et al. 2020. Role of plant growth-promoting bacteria in sustainable agriculture. *Biocatalysis and Agricultural Biotechnology* 30: 101842.

Pickett, J. A. 2013. Food security: intensification of agriculture is essential, for which current tools must be defended and new sustainable technologies invented. *Food and Energy Security* 2: 167–173.

Porat, R., V. Vinokur, B. Weiss, et al. 2003. Induction of resistance to Penicillium digitatum in grapefruit by β-aminobutyric acid. *European Journal of Plant Pathology* 109: 901–907.

Prasanna, R., A. Kanchan, B. Ramakrishnan, et al. 2016a. Cyanobacteria-based bioinoculants influence growth and yields by modulating the microbial communities favourably in the rhizospheres of maize hybrids. *European Journal of Soil Biology* 75: 15–23.

Prasanna, R., B. Ramakrishnan, K. Ranjan, et al. 2016b. Microbial inoculants with multifaceted traits suppress Rhizoctonia populations and promote plant growth in cotton. *Journal of Phytopathology* 164(-11–12): 1030–1042.

Prasanna, R., R. Sharma, T. Sharma, et al. 2013a. Soil fertility and establishment potential of inoculated cyanobacteria in rice crop grown under non-flooded conditions. *Paddy and Water Environment* 11(1): 175–183.

Prasanna, R., S. Babu, A. Rana, et al. 2013b. Evaluating the establishment and agronomic proficiency of cyanobacterial consortia as organic options in wheat–rice cropping sequence. *Experimental Agriculture* 49: 416–434.

Prasanna, R., S. Triveni, N. Bidyarani, et al. 2014. Evaluating the efficacy of cyanobacterial formulations and biofilmed inoculants for leguminous crops. *Archives of Agronomy and Soil Science* 60: 349–366.

Priya, H., R. Prasanna, B. Ramakrishnan, et al. 2015. Influence of cyanobacterial inoculation on the culturable microbiome and growth of rice. *Microbiological Research* 171: 78–89.

Pyakuryal, B. A. 2012. Feasibility study on the application of green technology for sustainable agriculture development: Assessing the policy impact in selected member countries of ESCAP-APCAEM. United Nations.

Rahman, M. A., M. M. Rahman, M. Ferdousi, et al. 2012. Use of culture filtrates of Trichoderma strains as a biological control agent against Colletotrichum capsici causing Anthracnose fruit rot disease of chili. *Journal of Biological & Environmental Sciences* 2: 9–18.

Rai, A., S. Rai, and A. Rakshit. 2013. Mycorrhiza-mediated phosphorus use efficiency in plants. *Environmental and Experimental Biology* 11: 107–117.

Rai, P. K., M. Singh, K. Anand, et al. 2020. Role and potential applications of plant growth-promoting rhizobacteria for sustainable agriculture. In *New and Future Developments in Microbial Biotechnology and Bioengineering*, edited by Gupta, V., 49–60. Elsevier, Amsterdam.

Rana, A., M. Joshi, R. Prasanna, et al. 2012. Biofortification of wheat through inoculation of plant growth promoting rhizobacteria and cyanobacteria. *European Journal of Soil Biology* 50: 118–126.

Rana, A., S. R. Kabi, S. Verma, et al. 2015. Prospecting plant growth promoting bacteria and cyanobacteria as options for enrichment of macro- and micronutrients in grains in rice-wheat cropping sequence. *Cogent Food and Agriculture* 1: 1.

Rashad, S., A. S. El-Hassanin, S. S. M. Mostafa, et al. 2019. Cyanobacteria cultivation using olive milling wastewater for bio-fertilization of celery plant. *Global Journal of Environmental Science and Management* 5: 167–174.

Riaz, M., M. Kamran, Y. Fang, et al. 2020. Boron supply alleviates cadmium toxicity in rice (Oryza sativa L.) by enhancing cadmium adsorption on cell wall and triggering antioxidant defense system in roots. *Chemosphere* 266: 128938.

Saffeullah, P., N. Nabi, S. Liaqat, et al. 2021. Organic agriculture: Principles, current status, and significance. In *Microbiota and Biofertilizers A Sustainable Continuum for Plant and Soil Health* edited by Hakeem, K. R., G. H. Dar, M. A. Mehmood, and R. A. Bhat, pp. 17–37. Springer, Cham.

Selvaraj, A., and K. Thangavel. 2021. Arbuscular mycorrhizal fungi: Potential plant protective agent against herbivorous insect and its importance in sustainable agriculture. In *Symbiotic Soil Microorganisms* edited by Shrivastava, N., S. Mahajan, and A. Varma, pp. 319–337. Springer, Cham.

Shahane, A. A., Y. S. Shivay, R. Prasanna, et al. 2019. Nitrogen nutrition and use efficiency in rice as influenced by crop establishment methods, cyanobacterial and phosphate solubilizing bacterial consortia and zinc fertilization. *Communications in Soil Science and Plant Analysis* 50(12): 1487–1499.

Shariatmadari, Z., H. Riahi, M. Seyed Hashtroudi, et al. 2013. Plant growth promoting cyanobacteria and their distribution in terrestrial habitats of Iran. *Soil Science and Plant Nutrition* 59(4): 535–547.

Sharma, I. P., A. K. Sharma, L. Prashad, et al. 2018. Natural bacterial cell-free extracts with powerful nematicidal activity on root-knot nematode. *Rhizosphere* 5: 67–70.

Sharma, I. P., and A. K. Sharma. 2020. Trichoderma–fusarium interactions: A biocontrol strategy to manage wilt. In *Trichoderma* edited by Sharma, A. K., and P. Sharma, pp. 167–185. Springer, Singapore.

Sharma, I. P., C. Kanta, T. Dwivedi, et al. 2020. Indigenous agricultural practices: A supreme key to maintaining biodiversity. In *Microbiological Advancements for Higher Altitude Agro-Ecosystems & Sustainability* edited by Goel, R., S. Ravindra, and D. C. Suyal, pp. 91–112. Springer, Singapore.

Sharma, I. P., S. Chandra, N. Kumar, et al. 2017. PGPR: heart of soil and their role in soil fertility. In *Agriculturally Important Microbes for Sustainable Agriculture* edited by Meena, V. S., P. K. Mishra, J. K. Bisht, et al., pp. 51–67. Springer, Singapore.

Swarnalakshmi, K., R. Prasanna, A. Kumar, et al. 2013. Evaluating the influence of novel cyanobacterial biofilmed biofertilizers on soil fertility and plant nutrition in wheat. *European Journal of Soil Biology* 55: 107–116.

Thajuddin, N., and G. Subramanian. 2005. Cyanobacterial biodiversity and potential applications in biotechnology. *Current Science* 89: 47–57.

Tian, T., A. Reverdy, Q. She, et al. 2020. The role of rhizodeposits in shaping rhizomicrobiome. *Environmental Microbiology Reports* 12(2): 160–172.

United Nation. 2018. https://www.unenvironment.org/news-and-stories/press-release/banking-sunshine-world-added-far-more-solar-fossil-fuel-power.

Vejan, P., R. Abdullah, T. Khadiran, et al. 2016. Role of plant growth promoting Rhizobacteria in agricultural sustainability-a review. *Molecules* 21: 1–17.

Verma, S., A. Adak, R. Prasanna, et al. 2016. Microbial priming elicits improved plant growth promotion and nutrient uptake in pea. *Israel Journal of Plant Sciences* 63(3): 191–207.

Vero, S., G. Garmendia, M. B. Gonzalez, et al. 2013. Evaluation of yeasts obtained from Antarctic soil samples as biocontrol agents for the management of postharvest diseases of apple (Malus×Domestic). *FEMS Yeast Research* 13: 189–199.

Waller, F., B. Achatz, H. Baltruschat, et al. 2005. The endophytic fungus Piriformospora indica reprograms barley to salt-stress tolerance, disease resistance, and higher yield. *Proceedings of the National Academy of Sciences USA* 102: 13386–13391.

Wang, Z., J. Shao, Y. Xu, et al. 2015. Genetic basis for geosmin production by the water bloom-forming cyanoacterium, Anabaena ucrainica. *Water* 7: 175–187.

Whitelaw, M. A. 1999. Growth promotion of plants inoculated with phosphate-solubilizing fungi. *Advances in Agronomy* 69: 99–151.

Whittinghill, L. J., and D. B. Rowe. 2012. The role of green roof technology in urban agriculture. *Renewable Agriculture and Food Systems* 27(4): 314–322.

Yuan, W. M., and D. L. Crawford. 1995. Characterization of Streptomyces lydicus WYEC108 as a potential biocontrol agent against fungal root and seed rots. *Applied and Environmental Microbiology* 61: 3119–3128.

Yuan, Z., A. Ali, P. Ruiz-Benito, et al. 2020. Above-and below-ground biodiversity jointly regulate temperate forest multifunctionality along a local-scale environmental gradient. *Journal of Ecology*. doi:10.1111/1365-2745.13378.

Zahra, Z., D. H. Choo, H. Lee, et al. 2020. Cyanobacteria: Review of current potentials and applications. *Environments* 7(2): 13.

4

Briquetting: A Technique for the Densification
of Biomass for Agriculture Waste

Garima Kumari and Sunita Rawat

CONTENTS

DOI: 10.1201/9781003175926-4

4.1 Introduction

The recent global scenario advises that energy consumption has increased as the community is facing energy crisis in this growing world. The primary cause is population growth, which is dependent on available energy resources for development, growth, and well-being (Sansaniwal et al. 2017). Ultimately, this is pressurizing the accessible sources, i.e. fossil fuels. Biomass plays a significant role in meeting the energy needs of people in developing countries. It is a well-known fact that the world is facing crucial problems such as global warming, deforestation, pollution and a shortage of fossil fuels, which has resulted in a rise in fuel prices. As a result, biomass energy is emerging as a sustainable and promising source of energy, as biomass production results in less or no net carbon dioxide aggregation in the atmosphere (Tumuluru et al. 2010a).

In this case, this is since the carbon dioxide released during the combustion process is compensated by the carbon dioxide used in photosynthesis. Other advantages of using biomass as an energy source include its infinite availability, ability to be processed locally, and environmental friendliness (Sinha et al. 2019). Agro residues, fruit waste, plant debris, wood waste, and other plant material are examples of readily accessible biomass sources. Furthermore, the disposal of these wastes continues to be a burden for farmers and commoners because they cause annoyance and are unsightly. As a result, using these agricultural residues in energy generation is an alternative and appreciated easy fix to this issue (Oladeji 2012). Biomass residues are employed in making briquettes, pellets, biogas, and bio fuel, among other things.

4.2 Densification and Its Need

The method of transforming biomass into a lightweight, combustible solid with a high density and calorific value is known as densification. Densification allows the density, porosity, particle size, and other physical dimensions of biomass to become more consistent and homogeneous, thus improving the thermal properties of the fuel.

The explicit burning of biomass resources to draw energy is a way old practice done in the country. Since direct burning of biomass residues is ineffective, rather than burning the unprocessed biomass resource in its entirety, which is dense and wet, the raw material should be processed and compacted to make transportation and handling easier. Apart from this, the raw material is rich in oxygen and low in carbon percentage, resulting in a low calorific value. As a result, it should be treated and compressed to maximize its fuel value. The major disadvantage behind using raw biomass directly for energy production is its low density, which ranges from 80–100 kg/m³ for agricultural straws and grasses to 150–200 kg/m³ for woody biomass, like woodchips (Sokhansanj and Fenton 2006; Mitchell et al. 2007; Tumuluru et al. 2010a; Kumar et al. 2016, 2019). The low bulk density of biomass retains the challenge of transportation, handling, storage, and use (Pallavi et al. 2013). This also impedes the co-firing process, as differences in bulk densities cause process interruption in coal co-firing. Furthermore, the advantages of using densified biomass include lower transportation costs due to increased energy density, convenient feeding, mechanical handling, and low dust production (Clarke and Preto 2011).

4.3 Briquetting

Briquetting is one of the various methods of densification. It is one of the feasible technologies that helps in converting the biomass energy (Wilaipon 2009). Briquetting has been practiced in several countries for many years. The biomass potential in India is estimated to be 61,000 MW, and the estimated employment generated by the industry is about 15.52 million (Sinha et al. 2019). Briquettes are made by putting residual mass under high pressure with or without a binder and compressing it into a compact form with a high bulk density, low moisture content, consistent size and shape, and a high ignition value. Briquettes usually have a diameter of more than 25 mm and a length of 150 mm, ranging from 80 to 150 mm. In saw dust cylinders compressed at high temperature, the density is 1.2–1.4 g/cm³ from loose biomass with a bulk density of 0.1–0.2 g/cm³ and moisture content of 10%–20%. Due to thermoplastic movement, the biomass particles form a self-bond between them during the briquetting process. The biomass becomes more compact as a consequence of this bond, ultimately resulting in compact briquettes (Tumuluru et al. 2010a). Briquettes come in a variety of shapes, from cylinders to rectangular or prismatic, with some having holes to aid ignition.

Briquetting is a viable and intriguing method of converting biomass into a fuel. Biomass is a clean and green approach for producing fuel that can be used for a variety of purposes. It is important to improve the attributes of feedstock by compacting them to increase their bulk density, making them easier for their storage, generation of electricity and transport.

4.4 Raw Materials for Briquetting

A variety of agro residues like sugarcane bagasse, barley, wheat straw, banana peel, canola, saw dust, tapioca waste, wood chips, coir pith, castor husk, pulses stalks, cotton sticks, mustard stalks, coffee husk, rice husk, bamboo dust, sander dust, groundnut shell, tamarind pods, coconut shell powder, tea waste, pine needle and wild grasses, and shrubs are most commonly used in India for briquetting (Wilaipon 2007; Oladeji et al. 2009; Tumuluru et al. 2015; Singh et al. 2020). These residues can be used singly or in combination. Raw materials are chosen for their ease of availability in the environment as well as their potential to densify and bond together. The compaction of these materials along with added binding agents results in briquetting (Adapa et al. 2002, 2009; Sokhansanj et al. 2006; Sanap et al. 2016; Aishwariya and Amasamani 2018).

4.5 Assessment of Quality Parameters Involved in Briquetting

Consumer utilisation is influenced by the quality characteristics of the densified biomass. The moisture content, ash content, particle size, flow characteristics, and other factors influence raw material selection. Biomass with up to 4% ash content is preferred for briquetting as it has materials with homogeneous granules that flow easily in bunkers, conveyors, and storage silos (Grover and Mishra 1996).

4.5.1 Moisture Content (%)

The high moisture content of the raw material can lead to microbial decay and spoilage. As a result, the preferred moisture range for briquettes varies depending on the raw material content. The optimum moisture content is 6%–15%, whereas the studies by Mani et al. (2006) on corn stover densification suggest that low moisture (5%–10%) is suitable for denser, more robust, and durable briquettes. They also concluded that moisture content greater than 15% and pressure greater than 15 MPa have a negative effect on the final product. The high moisture content creates complications in grinding and requires more energy for drying, as well as leading to swelling and cracks in the product (Grover and Mishra 1996).

4.5.2 Bulk Density (kg/m³ or g/cm³)

The most important factor of the final compact product, i.e., briquette, is its bulk density. The higher bulk density product is preferred as a fuel due to its high energy content and slow burning property (Kumar et al. 2009). The range from 1 to 1.6 g/cm³ is found to be an optimum value of bulk density that tends to affect the burning quality of the briquette.

4.5.3 Durability Index (%)

When the densified product has the ability to retain its physical characteristics like its original shape at the time of material handling, storage, and transportation, this is known as durability index of the product. Pre-heating and conditioning of the material with steam increases the durability of the material. Also, lignin and starch-like binding material having desirable moisture content help improve the durability of the product (Kaliyan and Morey 2009).

4.5.4 Percent Fine and Ash Content (%)

Transportation and storage of the densified briquettes cause its breakdown, which leads to the generation of fine particles. These fine particles are an obstacle in industries, especially where co-firing of fossil fuel takes place, as it can obstruct the process of co-firing (Tumuluru et al. 2010a). Other factors like low moisture content, slow rotational speed, and lesser binding content also lead to fine production. Generally, the widely accepted standard followed for the percent fine of densified biomass is 3.15 mm screen sieve.

Ash content is the percentage of content that remains after complete combustion of the material. The value of ash content may vary from 4% to 8%. The residue of biomass generally has much lower ash content, but their ashes contain higher concentration of alkaline minerals. These minerals liquefy during the heating process and condense on the tubes, thus impeding the process (Grover and Mishra 1996).

4.5.5 Calorific Value (MJ/kg)

The calorific value of the biomass majorly depends upon its particle size, the bulk density, temperature, and the type of pre-treatment, compaction pressure, type of binding agent used, etc. In general, the higher the bulk density, higher is the calorific value of the briquettes. Generally, the calorific value of briquettes ranges from 17 to 27 MJ/kg (Tumuluru et al. 2010b; Narra et al. 2010).

4.5.6 Flow Characteristics

The material should be granular and homogenous in order to have convenience in moving from bunkers to bunkers and in storage silos (Grover and Mishra 1996).

4.6 Bonding Mechanism in Biomass Briquetting

A pre-requisite for briquetting is to know about the physical and chemical properties of the biomass. Converting biomass into fuel entails a series of procedures that compact the loose raw material by reducing its volume and densifying it. According to Mani et al. (2002), biomass densification occurs in three phases. The energy is released due to inter-particle and particle-to-wall friction in the first step, as particles reorganize themselves to form a densely packed mass. Most of the particles maintain their properties and the energy is released. In the second stage, the particles are pressed against each other and undergo plastic and elastic deformation, considerably enhancing inter-particle contact. Van der Waals' electrostatic forces bind particles together. The density of the briquette exceeds the true density of the component in the third step at higher pressures due to a significant volume reduction.

When biomass is compacted under high pressure, the particles are forced to bond and adhere to one another, resulting in mechanical intertwining (Kaliyan and Morey 2009). This gives the briquettes their distinct shape and size. The quality of the densified product formed is determined by the firmness and stability of the bonds. Bond interlocking, adhesion-cohesion forces, and intermolecular forces between the particles are all used in biomass binding under high pressure. Fibres and heavy particles in biomass tend to stick together and form tight bonds. The system should be operated with optimal compression and shear forces to achieve this type of bonding mechanism.

Kaliyan and Morey (2010) investigated the densification of corn straw and switchgrass. They used a scanning electron microscope to detect the formation of solid bridges (SEM). The SEM revealed that the inter-particle bonding in the biomass was primarily due to concrete bridges formed by natural binders such as protein. Another method known as UV auto fluorescence was used to confirm this bridge formation. Furthermore, the studies discovered that moisture and temperature play an essential role in stimulating the naturally occurring binders in biomass. This ensures long-term inter-particle bonding.

4.7 Step-by-Step Process Involved in Biomass Briquetting

The process of briquetting involves a series of steps (Figures 4.1 and 4.2):

- Raw material collection
- Pre-treatment of raw material
- Densification
- Cooling and storage

4.7.1 Raw Material Collection

Briquetting is usually done with agro residues that can easily burn but are not in a specified shape or size.

4.7.2 Pre-treatment of Raw Material

Prior to densification, biomass is subjected to several treatments:

i. *Grinding*—Cutting, grinding, slicing, hammering, smashing, and chopping are used to reduce the size of the biomass until it is uniform and the desired size is reached (Solano et al. 2016). This produces a higher density product that uses minimal energy during the densification process (Clarke and Preto 2011)

ii. *Drying*—Low moisture content in the raw material gives the fuel its desired density and durability (Shaw and Tabil 2007). The raw material is dried in the light, open-air, or in solar dryers

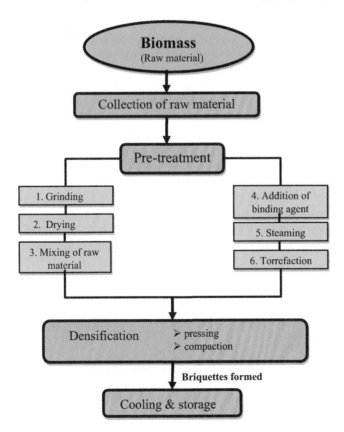

FIGURE 4.1 Flowchart of steps involved in briquetting.

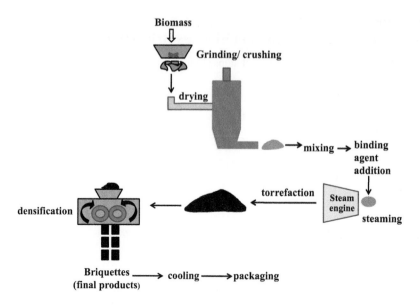

FIGURE 4.2 Process of biomass briquetting.

with heaters or hot air flow. Beyond the optimal moisture levels, the strength of the biomass and resilience are greatly influenced. As a result, one of the most significant phases in biomass densification is drying.

iii. *Mixing of raw materials*—This step is entirely optional and not necessary. To make briquettes, it is preferable to combine one or more raw materials. A classic example is when high-ash biomass is mixed with low-ash biomass. Likewise, low-energy content biomass can be mixed with high-energy content biomass. The proportion of raw materials used in the mixing should have a desirable compression and a high calorific value. This aids in achieving the desired quality, such as extended burning time and non-smoking and odourless briquettes (Oladeji 2015)

iv. *Addition of binding agent*—The natural binders present in biomass have a significant impact on its density and durability. The binding ability of the material increases as the protein and starch content increases. Binding agents such as vegetable oil, clay, starch, bitumen, resin, cooking oil, or wax are added to improve the binding ability of the biomass content. Corn stalks, on the other hand, have a high binding power (Kaliyan and Morey 2006).

v. *Steaming*—Prior to densification, steaming the biomass allows the natural binders to be released and activated (Kaliyan and Morey 2006).

vi. *Torrefaction*—It is a thermal method that involves slowly heating biomass in the absence of oxygen at temperatures between 280°C and 320°C (Mani 2008; Clarke and Preto 2011). This process releases more oxygen and hydrogen than carbon, increasing the calorific value of the product (Uslu et al. 2008). The resulting product is brittle and has improved grindability and fuel quality for combustion. It is also known as a mild version of slow pyrolysis that aims to dry and waterproof the biomass, making it resistant to moisture and biological attack.

$$\text{Biomass} \xrightarrow{\text{slow heating without } O_2} \text{Torrified product / biocoal} + CO + CO_2 + H_2 + \text{acetic acid}$$

4.7.3 Densification/Compaction

Hand pressing and compaction are two compression methods that can be used to compress biomass material to make it functional and transportable.

4.7.4 Cooling and Storage

The briquettes that result from compaction from the machines are extremely hot, reaching temperatures of over 100°C. The briquettes are then cooled and stored in a dry location until they are required again.

4.8 Techniques for Biomass Briquetting

Biomass, especially agro-residues, is processed using three types of techniques based on compaction (Wilaipon 2009):

4.8.1 Low-Pressure Compaction

In this process, a pressure as low as 5 MPa and room temperature are used, along with the addition of a binding agent to the residue. This briquetting technique is best for biomass having a dearth of lignin.

4.8.2 Medium-Pressure Compaction

Here, since the pressure is maintained between 5 and 100 MPa, little heat is generated; therefore, providing external heat is mandatory to melt the lignin content of the biomass. The use of a binding agent is no longer necessary after this.

4.8.3 High-Pressure Compaction

A high pressure of up to 100 MPa is applied in this technique, which in turn causes the temperature to rise to 200°C–250°C. This technique is advisable for high lignin content residues in which the high temperature and pressure cause fusion of lignin content. Therefore, there is no need of adding external binding agents to the biomass residue. The commonly used briquetting techniques include high-pressure compaction of the biomass residues.

4.9 Equipment for Biomass Briquetting

Based on various principles of operation, there are several briquetting machines available.

4.9.1 Screw Press/Screw Extruder

In 1945, Japan developed this technology entirely on its own. The biomass is dried in this process by moving hot air through it. The biomass is then heated in a heat exchanger to a temperature of 100°C–120°C. A revolving screw feeds the pre-heated biomass into a hopper, which then moves into a barrel. The barrel's own wall friction, the internal friction of the biomass material or residue, and the higher rotational speed of the screw around 600 rpm cause heating in the biomass as the temperature rises in this closed system. Suppose the closed system is unable to provide enough heat to bring the biomass material to a pseudo-plastic state for creamy discharge. In that case, heat is supplied to the extruder from the outside via band or tape heaters (Grover and Mishra 1996). The heated biomass material is then forced from the heater towards the extrusion die where the briquettes or pellets of desired shapes are formed. The screw press machines work by using a feeder screw to feed material into a die repeatedly (Figure 4.3 a).

Biomass processing in screw press involves (Grover and Mishra, 1996) the following steps:

- *1st stage*—During this stage, the greatest amount of energy is needed to overpower particle friction. Before reaching the compression zone, the biomass is mildly compressed to compact it.
- *2nd stage (Initial compression)*—The biomass material becomes soft and loses its elasticity in the compression zone due to the high temperature. The particles come closer to each other

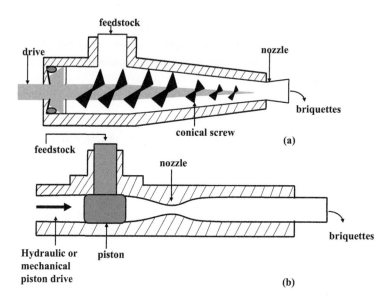

FIGURE 4.3 Briquetting equipment: (a) screw press and (b) piston press.

and form a link, resulting in inter-linking of the particles, and the friction that results helps the biomass material absorb energy evenly.

- *3rd stage (Final compression)*—Here, the compression on the biomass material is increased further as it reaches the tapered die, where the normal temperature causes moisture to evaporate.
- *Final stage*—In this stage, the compaction is released while the steam is removed simultaneously, resulting in a normalization of pressure throughout the system and the formation of homogeneous briquettes.

    Screw press products are denser and heavier, with concentric holes that increase the surface area of the component, increasing its ignition potential. The briquettes produced are of higher quality and do not crumble as easily as briquettes made with a piston press. The three main kinds of screw presses are conical screw presses, screw presses with heated dies, and screw presses without heated dies.

4.9.2 Piston Ram Press

Ram and die technology is another name for this. It is one of the most widely used high-pressure briquetting systems. The raw material is fed into a cylinder, which moves towards a tapered die when compressed by a piston, either mechanically or with the assistance of a hydraulic piston. Due to frictional force, the compressed material is fully heated at this point, and the lignin content of the material begins to flow. As a natural binder, this process aids in the binding of the compressed material. When the lignin comes from the die, it stiffens and compacts to form briquettes that range in size from 10 to 30 cm. The final product is completely solid in nature. The output of piston ram press machines ranges from 25 to 1,800 kg/h, depending on the properties of the raw material, its nature, and the canal diameter of the press. There are two kinds of piston presses: mechanical and hydraulic presses (Figure 4.3b) (Grover and Mishra 1996).

4.9.3 Mechanical Piston Press

Work done in Germany in the 1930s was the key source of inspiration for modern mechanical piston presses. During the Second World War, it was further developed in Switzerland. The raw material is pushed towards the die by a reciprocating piston (Tumuluru et al. 2011, Solano et al. 2016). The machine consists of an electric motor with a coupling belt. The mechanical machines are large and heavy and range from 0.3 to 0.45 t/h. They produce sturdy briquettes.

4.9.4 Hydraulic Piston Press

It works on the same principle as a mechanical piston press in reality. The only difference is that energy is transmitted to the piston through a high-pressure hydraulic oil system from an electric motor. Because the movement of the cylinder is slower than mechanical processes, the output of a hydraulic press is lower. The pressure is limited to 40–135 kg/h; therefore, briquettes have a bulk density less than 1,000 kg/m^3 (Tumuluru et al. 2011). As a result, the machines are small and light. The machine's entire operation is controlled by a programme that can be changed as needed.

4.9.5 Hand Press/Manual Press Briquette Machine

Hand presses or manually operated briquetting approaches are used to meet the domestic fuel demand in rural areas. In rural areas, dry leaves, groundnut shells, rice husk, as well as paper and other readily available materials, are used to make briquettes. Before being mixed with a binding agent, the raw material is sun-dried in the open air. The hand compressed briquetting machine is a level-ground machine. It consists of a steering arm that serves as a handle for rotating the screw rod. A slurry or raw material is fed through a compression metal plate, and the water squeezing out of the raw material is drained through a drain pipe. A stop plate, pressure releasing screw, bottom punch plate, stopping plate, and other briquette machine components are included. The hand-pressed briquetting machine measures 114 cm in height and 47 cm in width and weighs approximately 16.7 kg. The raw material is fed onto the compression plate, and the screw rod is rotated clockwise, applying pressure to it. The majority of the water is squeezed out this way. Depending on the thickness and water content of the material, the squeezing process is repeated 3–4 times. Finally, the briquettes are collected and dried in the sun before being stored for future use. The moisture content and bulk density of the so-formed briquettes should be maintained. The moisture content and bulk density of the briquettes that have been produced should be maintained (Madhava et al. 2012).

4.9.6 Beehive/Honeycomb Briquettes

In this process, agricultural residues, leaves, twigs, and branches should all be used as raw materials. The raw material is partially carbonized or pyrolyzed to produce these briquettes. After pyrolysis, the char is mixed with binders like clay and water in a 70:30 or 60:40 ratio. The mixture should be kept overnight for best results. The mixture is then forced through a cylindrical-shaped special mould, which produces briquettes that resemble a beehive or honeycomb, hence the name honeycomb or beehive briquettes. Following that, the briquettes are sun-dried for 2–3 weeks. The briquettes are then used as household fuel, and they are ignited with paper or dry wood (Nienhuys 2011).

4.9.7 Roller Press

The material is passed through two closed rotating rollers of equal diameter to produce briquettes in this process. The material is fed into the hopper and then pushed down into the gap by the machine's design. These rollers are now rotated in the opposite direction, compressing the material from the other side, causing it to move. The diameter of the roller, the gap between the rollers, and the shape of the die all influence the final product. Briquettes with a bulk density of 450–550 kg m^{-3} are produced (Yehia 2007; Kaliyan et al. 2009)

4.10 Essentialities for the Production of Biomass Briquettes

Briquette production is a long and time-consuming process that necessitates certain setup conditions. The following are the basic requirements for establishing a biomass briquetting plant:

1. *Land requirement*—A minimum of 1 acre of land is required to set up a briquette production plant to store the raw materials and the finished product.

2. *Raw materials*—Raw materials must be readily available in order for briquette production to be profitable.

3. *Drying facility*—Because the raw materials widely used have high moisture content, a proper drying facility should be available. As a result, equipment such as hot air generators, solar driers, and heaters are needed to reduce the moisture content to a manageable level for briquette production.

4. *Shredding machine*—To finely chop the agro residues for briquetting, a minimum of a 5 hp motor shredding machine is required.

5. *Briquetting machine*—In order to make briquettes, a high-pressure hydraulic piston press machine with a 50-hp motor is required. Screw press extruders, roller presses, and other briquetting machines are examples.

4.11 Advantages of Briquetting

Briquetting offers following advantages:

- *More proficient*—Briquettes, when compared to coal, have a better thermal effect and contain less ash. It produces a great deal of heat and lasts much longer than any other fuel. This is because of their low moisture content and bulk density.
- *Renewable*—Briquettes are made from readily available and renewable organic agricultural raw materials.
- *Easy preparation*—Briquettes are now manufactured commercially, but they can also be made at home.
- *Concentrated*—Briquettes are created by compressing volatile residues, resulting in compact, hard, and thick briquettes. As a result, they deliver a concentrated form of energy with a bulk density of approximately $800\,kg/m^3$.
- *Gradual burning*—Briquettes burn for a longer time due to their compactness.
- *Less pollution*—They produce less ash and contain no soot, smoke, or carbon deposits.
- *Easy transportation and storage*—They are simple to transport since the volume of biomass material is reduced by ten times, making them easy to store and transport. The simple cylindrical, cubical, and rectangular blocks are easy to manage and pack for transport.
- *Economical*—Briquettes are less expensive because they are made from domestic plant waste, and therefore sold at very low prices.

4.12 Limitations of Briquetting

Apart from being an excellent fuel replacement with numerous advantages, briquettes and the briquetting process have a number of disadvantages:

- Briquettes are only used in solid form, implying that they have no place in internal combustion engines.
- Because briquettes cannot withstand direct contact with water, adequate storage space is required to prevent water contact.
- Briquettes have low burning capacity per unit volume when compared to coal.
- Briquettes do not reach temperatures greater than 1,000°C as a result of their low carbon content, which is insufficient for industrial use.
- The screw in the briquetting machine has a finite life, breaking down after 3–4 hours of continuous use and being unrepairable after the tenth use. As a result, repairing it is a significant challenge in briquetting technology.

4.13 Applications of Biomass Briquettes

Briquettes have a wide variety of application in industrial, household, and commercial sectors (Oladeji 2015). Due to the high cost and shortage of charcoal, firewood, and other fuels, users are looking for a more cost-effective alternative (Table 4.1).

4.14 Biomass Briquetting in Present Scenario

Briquettes are in higher demand in developing countries, where basic necessities, including cooking fuel are scarce. On the other hand, briquettes are often used in developed countries for industrial purposes, such as electricity production from boilers. The biomass fuel industry is typically more advanced in developed countries at the moment. As environmental regulations become more stringent in the coming years, the Asia-Pacific region's biomass fuel market is expected to expand at the fastest pace. Government policies and incentives play a significant role in driving demand for industrial biomass briquettes and pellets for energy production. According to Fior Markets' Global Briquette Market Growth report, the briquette market is expected to grow at a 7.8% Compound Annual Growth Rate (CAGR) over the next 5 years.

India is a predominantly agricultural nation, with nearly 40% of the population employed in agriculture. As a developing country, it has limited fuel and energy resources. Because of their high demand, these resources are non-renewable and quickly depleting. To meet the need for these natural resources, the government is searching for environmentally sustainable and advantageous alternatives. Briquettes have a strong consumer demand due to their environmental benefits. These are an outstanding alternative to non-renewable fossil fuels. The current market situation for these briquettes in India is very strong due to their high demand and use. Briquettes are used because they protect the atmosphere from a range of toxins, conserve natural resources, and assist in waste management. Despite some setbacks, India is the only country where the briquetting industry is rapidly expanding (Reddy 2019). With fewer achievements and promotional campaigns, a growing number of entrepreneurs are now investing in biomass briquetting with confidence. Both national and international agencies have started projects and partnerships to upgrade the existing briquetting technology in India (Current Market Scenario of Briquettes in India 2014).

4.15 Challenges

As previously mentioned, briquetting technology is more advanced in developed countries as in developing countries. The production and utilization of briquettes in developed countries occur on a large scale compared to that in developing countries like India. The expansion of biomass densification is largely

TABLE 4.1

Application of Briquettes

S. No.	Sector	Application
1.	Gasification	Fuel for gasifiers to produce electricity
2.	Food processing industries	Distilleries, bakeries, drying
3.	Industrial boilers	Steam and heat generation
4.	Ceramic industries	Firing the furnaces of brick kilns, pottery, tiles
5.	Poultry	Incubating the chicks
6.	Agriculture	Oil milling, tea drying, tobacco curing, heating green houses and nurseries
7.	Textile processing industries	Drying, dyeing and bleaching purpose
8.	Charcoal production	Initiating pyrolysis
9.	Domestic purpose	Cooking, heating

dependent on components such as residue availability, suitable technology, capital requirements, customer knowledge, environmental impacts, and the briquette market (Felfli et al. 2011).

Residue availability is not an issue in developing countries like India, where biomass plants are primarily focused on agricultural wastes; however, optimizing the chemical and mechanical procedure needed for most of the feedstock remains a challenge. Regarding rural areas, where power is scarce, a suitable method of pre-processing would require the least amount of energy input. The majority of technologies for producing high-quality briquettes, as tested, are costly and require a lot of resources (Kpalo et al. 2020).

Briquette production on a large scale would necessitate a major capital expenditure. This poses a barrier to growing biomass densification. To encourage further investment in areas with limited financial resources and high energy input, efforts should be focused on creating more convenient, cost-effective, handy, and energy-efficient technologies at different scales. Despite their long life, few people are familiar with briquettes and how they vary from firewood and normal charcoal. Consumers in rural towns, in particular, have not yet been made aware of evidence of possible fuel savings and health benefits from the use of briquettes. Some carbonised briquettes have a longer burn time than charcoal, which is an advantage that a potential buyer might not notice when faced with a higher initial price. The commodity is unlikely to sell without adequate marketing and also costs more per unit weight than charcoal. It is possible to boost the livelihoods of millions of people, increase natural carbon sequestration, and help communities' ability to respond to climate change challenges by replacing wood-based charcoal with renewable fuels. Since cultural practices and household appliances depend on the use of charcoal, producing similar fuels from renewable sources such as agricultural waste is the most effective way to address the issues posed by non-durable wood charcoal (Kpalo et al. 2020). Finally, the demand for briquettes persists, and so does their widespread use as a biomass and biofuel alternative. However, in order to realize its full potential, the obstacles mentioned above must be addressed.

4.16 Conclusion

Biomass densification refers to a group of methods for converting biomass into fuel. Briquetting is one of the biomass densification technologies that improves the handling characteristics of materials for transport and their storage. The aim of briquetting raw materials is to increase the value of an existing product. Densification increases the volumetric calorific value of a fuel, lowers transportation costs, and can help improve the fuel situation in rural areas; thus, this technology can help extend the use of biomass in energy production. The purpose of agglomerating residues is to make them denser so that they can be used in energy production. Briquetting technology, which uses a variety of waste materials, should be promoted as an alternative to addressing modern-day energy issues because it is an efficient, cost-effective, and simple-to-implement technology. Appropriate briquetting equipment suitable for local communities must be developed for biomass to make a significant impact as a fuel for rural communities.

REFERENCES

Adapa, P., L. Tabil, and G.J. Schoenau. 2009. Compression characteristics of selected ground agricultural biomass. *Agricultural Engineering International: CIGR Journal* 11: 1–19.

Adapa, P., L. Tabil, G. Schoenau, S. Sokhansanj and J. Bucko. 2002. Compression and compaction behavior of fractionated alfalfa grinds. *ASAF ICSAE Proceedings*.

Aishwariya, S. and S. Amsamani. 2018. Exploring the potentialities and future of biomass briquettes technology for sustainable energy. *Innovative Energy and Research* 7: 221.

Clarke, S. and F. Preto. 2011. *Biomass Densification for Energy Production*. ON, Canada: Ontario Ministry of Agriculture, Food and Rural Affairs.

Current Market Scenario of Briquettes in India. 2014. http://briquettingmachine.weebly.com/blog/current-market-scenario-of-briquettes-in-india (Accessed: January, 2021).

Felfli, F.F., J.D. Rocha, D. Filippetto, C.A. Luengo and W.A. Pippo. 2011. Biomass briquetting and its perspectives in Brazil. *Biomass and Bioenergy* 35(1):236–242.

Fior Markets. 2019. Global Briquette Market Growth 2019–2024- Fior Markets.

Grover, P.D. and S.K. Mishra. 1996. *Biomass Briquetting: Technology and Practices* (Vol. 46). Bangkok, Thailand: Food and Agriculture Organization of the United Nations.

Kaliyan, N. and R.V. Morey. 2009a. Densification characteristics of corn stover and switchgrass. *Transactions of the ASABE* 52(3):907–920.

Kaliyan, N. and R.V. Morey. 2006. Factors affecting strength and durability of densified products. In *2006 ASAE Annual Meeting* (p. 1). American Society of Agricultural and Biological Engineers. .

Kaliyan, N. and R.V. Morey. 2010. Natural binders and solid bridge type binding mechanisms in briquettes and pellets made from corn stover and switchgrass. *Bioresource Technology* 101(3):1082–1090.

Kaliyan, N., R.V. Morey, M.D. White and A. Doering. 2009. Roll press briquetting and pelleting of corn stover and switchgrass. *Transactions of the ASABE* 52(2):543–555.

Kpalo, S.Y., M.F. Zainuddin, L.A. Manaf and A.M. Roslan. 2020. A review of technical and economic aspects of biomass briquetting. *Sustainability* 12(11): 4609.

Kumar N., N. Jeena, and H. Singh 2019. Elevated temperature modulates rice pollen structure: A study from foothill Himalayan Agro-ecosystem in India. *3Biotech* 9: 175.

Kumar, N., S.C. Shankhdhar and D. Shankhdhar 2016. Impact of elevated temperature on antioxidant activity and membrane stability in different genotypes of rice (*Oryza sativa* L.). *Indian Journal of Plant Physiology* 21(1): 37–43.

Kumar, R., N. Chandrashekar and K.K. Pandey. 2009. Fuel properties and combustion characteristics of Lantana camara and Eupatorium spp. *Current Science* 97: 930–935.

Madhava, M., B.V.S. Prasad, Y. Koushik, K.R. Babu and R. Srihari. 2012. Performance evaluation of a hand operated compression type briquetting machine. *Journal of Agricultural Engineering* 49(2):46–49.

Mani, S. 2008, March. Recent development in biomass densification technology. In *Annual Conference, Institute of Biological Engineering, Biological and Agricultural Engineering Department, University of Georgia.*

Mani, S., L.G. Tabil and S. Sokhansanj. 2002. Compaction behavior of some biomass grinds. AIC Paper (02–305).

Mani, S., L.G. Tabil and S. Sokhansanj. 2006. Effects of compressive force, particle size and moisture content on mechanical properties of biomass pellets from grasses. *Biomass and Bioenergy* 30(7):648–654.

Mitchell, P., J. Kiel, B. Livingston and G. Dupont-Roc. 2007. Torrefied biomass: A foresighting study into the business case for pellets from torrefied biomass as a new solid fuel. *All Energy* 24:2007.

Narra, S., Y. Tao, C. Glaser, H.J. Gusovius and P. Ay. 2010. Increasing the calorific value of rye straw pellets with biogenous and fossil fuel additives. *Energy & Fuels* 24(9):5228–5234.

Nienhuys, I.S. 2011. The beehive charcoal briquette stove in the Khumbu Region, Nepal.

Oladeji, J.T. 2012. Comparative study of briquetting of few selected agro-residues commonly found in Nigeria. *The Pacific Journal of Science and Technology* 13(2):80–86.

Oladeji, J.T. 2015. Theoretical aspects of biomass briquetting: A review study. *Journal of Energy Technologies and Policy* 5(3):72–81.

Oladeji, J.T., C.C. Enweremadu and E.O. Olafimihan. 2009. Conversion of agricultural wastes into biomass briquettes. *IJAAAR* 5(2):116–123.

Pallavi, H.V., S. Srikantaswamy, B.M. Kiran, D.R. Vyshnavi and C.A. Ashwin. 2013. Briquetting agricultural waste as an energy source. *Journal of Environmental Science, Computer Science and Engineering & Technology* 2(1):160–172.

Reddy, J. June 19, 2019. Briquetting Process, Techniques, Uses, Briquetting Types. https://www.agrifarming.in/briquetting-process-techniques-uses-briquetting-types (Accessed: February, 2021).

Sanap, R., M. Nalawade, J. Shende and P. Patil. 2016. Automatic Screw Press Briquette Making Machine. *International Journal of Novel Research in Electrical and Mechanical Engineering* 3(1):19–23.

Sansaniwal, S.K., K. Pal, M.A. Rosen and S.K. Tyagi. 2017. Recent advances in the development of biomass gasification technology: A comprehensive review. *Renewable and Sustainable Energy Reviews* 72:363–384.

Shaw, M.D. and L. Tabil. 2007. Compression, relaxation, and adhesion properties of selected biomass grinds. *Agricultural Engineering International: CIGR Journal* 9: 1–16.

Singh, H., M. Yadav, N. Kumar, A. Kumar and M. Kumar 2020. Assessing adaptation and mitigation potential of roadside trees under the influence of vehicular emissions: A case study of *Grevillea robusta* and *Mangifera indica* planted in an urban city of India. *PLoS ONE* 15(1): e0227380.

Sinha, Y., P. Sahu, S. Biswas, S. Manikpuri and M. Lathiya. 2019. Development of manually operated piston press type briquetting machine. *International Journal of Current Microbiology and Applied Sciences* 8(12):2932–2939.

Sokhansanj, S. and J. Fenton. 2006. *Cost benefit of biomass supply and pre-processing*. BIOCAP Canada Foundation.

Sokhansanj, S., S. Mani, M. Stumborg., R. Samson and J. Fenton. 2006. Production and distribution of cereal straw on the Canadian prairies. *Canadian Biosystems Engineering* 48: 3.

Solano, D., P. Vinyes and P. Arranz. 2016. Biomass briquetting process. *UNDP-CEDRO Publication: Beirut, Lebanon.*

Tumuluru, J.S., C.T. Wright, J.R. Hess and K.L. Kenney. 2011. A review of biomass densification systems to develop uniform feedstock commodities for bioenergy application. *Biofuels, Bioproducts and Biorefining* 5(6):683–707.

Tumuluru, J.S., C.T. Wright, K.L. Kenney and R.J. Hess. 2010a. A technical review on biomass processing: densification, preprocessing, modeling and optimization. In *2010 Pittsburgh, Pennsylvania, June 20-June 23, 2010* (p. 1). American Society of Agricultural and Biological Engineers.

Tumuluru, J.S., L.G. Tabil, Y. Song, K.L. Iroba and V. Meda. 2015. Impact of process conditions on the density and durability of wheat, oat, canola, and barley straw briquettes. *Bioenergy Research* 8(1):388–401.

Tumuluru, J.S., S. Sokhansanj, C.J. Lim, T. Bi, A. Lau, S. Melin, T. Sowlati and E. Oveisi. 2010b. Quality of wood pellets produced in British Columbia for export. *Applied Engineering in Agriculture* 26(6):1013–1020.

Uslu, A., A.P. Faaij and P.C. Bergman. 2008. Pre-treatment technologies, and their effect on international bioenergy supply chain logistics. Techno-economic evaluation of torrefaction, fast pyrolysis and pelletisation. *Energy* 33(8):1206–1223.

Wilaipon, P. 2007. Physical characteristics of maize cob briquette under moderate die pressure. *American Journal of Applied Sciences* 4(12):995–998.

Wilaipon, P. 2009. The effects of briquetting pressure on banana-peel briquette and the banana waste in Northern Thailand. *American Journal of Applied Sciences* 6(1):167.

Yehia, K.A. 2007. Estimation of roll press design parameters based on the assessment of a particular nip region. *Powder Technology* 177(3):148–153.

5

Perspective of Agro-Based Bioenergy for Environmental Sustainability and Economic Development

Dipti Singh, Manali Singh, Ishwar Prakash Sharma, Deepika Gabba, Upasana Gola, Neha Suyal, Nasib Singh, Puneet Negi, Narendra Kumar, Krishna Giri, Ravindra Soni, and Deep Chandra Suyal

CONTENTS

5.1 Introduction

Biofuels can be obtained from any biomass like plants, algae, or animal waste. Due to the global increase in the demand for energy and the recent instability in global oil prices, biofuel industries are attracting significant interest. The primary energy source is currently crude oil, which is

DOI: 10.1201/9781003175926-5

considered unsustainable, while world energy consumption is estimated to grow by 57% by 2030. In addition to concerns related to sustainability and economics, there has also been an increase in the environmental impact of petroleum being the sole source of energy. The burning of crude oil is presently raising the emissions of harmful greenhouse gases in the atmosphere, thus enhancing air pollution. Therefore, concerns related to environmental issues and the rising price of crude oil have promoted research into alternative energy. Microalgal biofuel is known as a third-generation biofuel with the option of potential generation energy sources (Table 5.1). This fuel can help in more reduction of GHGs emissions in our environment. Biofuels include mainly biodiesel, bioethanol, biogas, and biomass. Renewable biofuels can displace a part of the fossil fuels because biofuels have the potential to reduce emissions of greenhouse gases and maintain environmental superiority (Hajjari et al. 2017).

Compared to others like soybean or other terrestrial-producing crops, microalgae biomass can build up (100 times) more oil components. Microalgae can produce 5,000–20,000-gallon oil yield per acre per year, which is 7–30 times more than any best crops. It can easily grow entirely dependent on water, CO_2 and sunlight.

TABLE 5.1

Biofuels and Their Characteristics

S. No.	Biofuel Name	Source	Characteristic Features	Drawbacks
Primary Biofuels				
1.	Pellets	Wood, agricultural waste, easier to use, need less storage space.	high energy content, high density, low moisture content	CO production, higher land use is required, forest destruction
2.	Woodchips	Waste wood, Pulpwood; forests, agricultural waste, easier to use	Moisture retention, temperature moderation, high energy value	Production of harmful gases, destruction of forests, large space requirement
3.	Landfill gases	Natural by-product of the decomposition of organic material	Reduction in greenhouse gases, less odour, health and safety benefits	May contaminate soil and water, may affect human health
Secondary Biofuels				
First generation (Figure 5.1)				
1.	Ethanol	Biomass, grasses, trees, agricultural residues	High octane level, lesser particle emission, safer for the health, engine modification is not required	Cause corrosion, engine burns, requires large cropland space, costly option
2.	Biodiesel	Sunflower oil, palm oil, rapeseed oil, Waste animal fats,	Higher octane value, long-lasting, improved engine efficiency	NO release, pollutant, less appropriate during low-temperature transport
3.	Bioethers	Partial biomass oxidation, Dehydration of bio-based alcohol	Higher combustion and emission efficiency, gasoline additive, Non-toxic, sulphur free	Expensive, lower lubricity and viscosity
Second generation (Fig)				
1.	Waste vegetable oil	Oil crops	Less environmental pollution, cost-effective	Diesel engine can be damaged, difficult to collect

(*Continued*)

TABLE 5.1 (*Continued*)

Biofuels and Their Characteristics

S. No.	Biofuel Name	Source	Characteristic Features	Drawbacks
2.	Butanol	Biomass, Corn grains	Farmer friendly, High calorific value, more octane number gasoline additive, no corrosion	May be toxic, incompatible with some fuels, require larger area to grow the crops
Third generation				
1.	Biodiesel from microalgae	Algal biomass	High oil content, Land is not required, CO_2 removal	Biomass production is low, costly process, lesser lipid content
2.	Algal hydrogen	Algal biomass	Less expensive, several carbon sources can be used, efficient	Completely dependent upon hydrogenase enzyme system. This activity may reduce under environmental fluctuations
Fourth-generation				
1.	Electrofuels	CO_2 and H_2O along with electricity	Efficient, provide stability to the grids, compatible with several other energy sources	Very expensive
2.	Solarfuels	Sun	Clean source, Renewable source, cost-effective, environment-friendly	Storage power, Requires large space, dependent on weather conditions

5.2 Biomass: A Renewable Energy Resource

Generally, the biomass refers to all organic matter in the nature that originated from the plant, animal, and microbes. For human civilization, several ways to produce energy from biomass exist as the generation of heat from burning biomass for electricity, thermal power, biofuel, etc. It has been estimated that in United States nearly 680 million tons of biomass resources will be able to harness up to 2030, which was sufficiently observed to be suitable to generate 10 billion gallons of ethanol or energy of 166 billion kilowatt-hours electricity. Moreover, the International Energy Agency (IEA) has reported that a total of 17% of the global population is still without electricity, with an estimation of 1.2 billion people. The main purpose of renewable energy sources is to limit greenhouse gas emissions, diversify energy supply, and reduce dependence on unreliable and volatile fossil fuel markets. For the generation of natural energy, we need renewable resources. Various major natural biomass resources used as renewable energy are as follows.

5.2.1 Wood Resources from Natural Forests and Commercial Plantations

Among all the natural biomasses, wood is one of the largest biomass energy sources. The wood biomass generally obtains from forests and also can be easily grown outside the forest. Since the discovery of fire, wood has been used for cooking and heating, which is used by more than 2 billion people. Hence, it is the most important source of bioenergy. Wood is an important natural fuel used by human beings for thousands of years. From the global energy supply, 18% is renewable energy; nearly 13% of this is attributed to traditional biomass. Wood fuel is environmentally friendly and has a low-risk energy carrier and ensures safe handling and storage. Commercially in various developing countries, it is applicable in fish

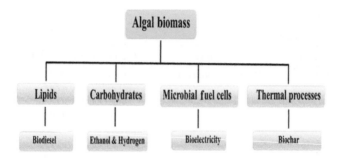

FIGURE 5.1   Biofuels derived from algae.

drying, tobacco curing, brick baking, etc. Biofuels derived from wood biomass include firewood, wood shavings, pellets, fruit stones, and nutshells. Among these, the cut and chopped firewood are useful in domestic applications such as direct burning. The woody parts of fruit, such as fruit stones, husk, seeds, etc., are also used for domestic energy due to solid biofuel. In various studies, mango stone, peanut shell, and sunflower seed husk were reported for high energy potential, which has a higher heating value than other commercial biofuels (Figure 5.1).

5.2.2 Agricultural Residues and Bioenergy Production

Agricultural waste such as crops, grasses, and tree residues is the source of cellulosic biomass. Generally, we consider crop residues as waste while providing important environmental processes to the ecosystem in a real sense. These residues are varying in particle size, moisture content, bulk density, etc., which are generally fibrous and low in nitrogen, depending on the geography or location in which they occur. These field wastages are used as fertilizer and fodder for livestock, but it is beneficial to use for energy due to carbon storage during growth from the air, which is essential for carbon sequestering and power generation. Among all residues, corn residues, leaves, and stalks leftover in the field after corn harvesting are very useful for ethanol production. Such commercial utilization of residues can reduce the use of natural gas and electricity. The residues contain a high amount of lingo-cellulosic biomass, which undergo slow degrading that causes long-term degradation of soil fertility and ultimately reduces soil organic matter, leading to the rapid promotion of CO_2 emissions (Kumar et al. 2015, 2109; Azadbakht et al. 2021). Lignin contains a much higher energy content as compared with cellulosic biomass. In India, rice grain husk has been shown to gasify in small-scale eco-friendly electricity generation units; such a renewable energy model must be used globally to process energy production. A huge amount of endocarp from drupe fruits is mostly thrown out after processing that is predominantly made up of lignin, with the weight of more than 50% wt/wt. Microbes are the most effective organism on the earth that convert glucose to ethanol by biological catalysts. Ethanol is one of the most important products after fermentation. However, some other substances such as hydrogen, methanol, and succinic acid are also generated. The major fermentation substrates are hexoses, while other carbohydrates such as pentose, glycerol, and other hydrocarbons are converted into ethanol.

5.2.3 Agro-Industrial Wastes Utilization for Bioenergy Requirements

A huge amount of organic residues is produced every year through the food processing industries, which can be utilized for the different energy sources such as agricultural residues that are indigenous, non-polluting, and virtually inexhaustible. Wastage from fruits, vegetables, and root crop generated after processing might be a good energy source. This production increases every year so the wastage percentage will also increase from them. The waste from food industries have a high value of biological oxygen demand (BOD) and Chemical oxygen demand (COD) that harm the environment; hence, we need to consume them by converting them into energy because they contain a huge amount of organic compounds. Especially in oil industries, huge amount of processed residues are left after oil extraction and known as oil cake. These cakes contain a high concentration of fat, oil, and suspended and dissolve solids that can be converted into energy after fermentation processes. With the help of biological resources, a large amount of ethanol has been produced in recent years. The fruit wastes are good for the production of single-cell protein that are good sources of energy because these wastes contain a high amount of proteins. The industrial-based wastes consist of valuable nutrients for the proper growth of microbes, leading to the production of different enzymes after fermentation. For enzyme production, these wastes can be used as a raw material in fermentation for desirable enzyme production through responsive microbes.

5.2.4 Livestock Waste for Bioenergy Production

Livestock waste is a significant potential source of renewable energy that can generate energy via an anaerobic digestion system, *i.e.* bio-refinery unit. Biogas is one of the best suitable and most straightforward renewable energy sources derived from sewage, liquid manure from animals, and organic agricultural wastage.

5.2.5 Municipal Solid Waste

Various human activities are responsible for the municipal solid waste, which need to be managed due to its pollution activity towards the environmental system. It is now a troubling problem for both the government and the public. Based on the World Bank Report 2012, solid waste will increase from 1.3 billion to 2.5 billion tons/year by 2025. In India, the urban area produces 120,000 tons/day/capita. The energy generation from this municipal waste is one of the best ways to reduce environmental pollution for power generation or industrial use. For energy generation, the solid waste is first broken down into small pellets, wood briquettes, and wood chips by the various processes such as physical, thermal, and biological methods that can be used for the part of energy generation in industries and domestic areas through biogas generation, methanol production, etc.

5.2.6 Bamboo-Based Bioenergy Resources

Bamboo is a good source of energy because it has low ash content and alkali index. Bamboo has a very high heat value and low moisture content (8%–23%) compared with any other plants. Pyrolysis is a thermochemical conversion method in which biomass degradation is done at a moderate-to-high temperature in the absence of oxygen that is useful to convert biomass to solid fuels (charcoals), liquid fuels, and gas (syngas) (Xing et al. 2021). Biochemical conversion is one of the best pathways for energy production. In this method, various strains of micro-organisms are utilized for different biofuel products; fermentation is the basic principle for this method in which the sugar or other substances present in biomass convert into ethanol, methane, and other fuels.

5.2.7 Microbial-Based Bioenergy Resources

5.2.7.1 Role of Microbial Bioresources in Bioenergy Generation

Mankind has, for most of its survival, depend on renewable energy resources like wood, water wheels, windmills, and animals such as oxen and horses. Biomass is an organic material acquired from living organisms, including plants, animals, algae, and microorganisms. Biofuels viz. biodiesel, hydrogen, bioethanol, and jet fuel are primarily algal and plant origin (Singh and Gonzales-Calienes 2021). We can categorize biofuel-producing organisms into two separate groups: energy harvesting group and energy transforming group. Photosynthetic organisms like plants, algae, cyanobacteria, and photosynthetic bacteria are kept in the first group, while heterotrophic organisms like animals, bacteria, and yeast belong to the second group.

Presently, the biofuel market is dominated by bioethanol and biodiesel that accounts for 90% of the production. This comes from the heterotrophic group as yeasts are responsible for bioethanol production. However, the traditional bioethanol producing sources, viz. maize, sugarbeet, etc., are now discouraged to protect the global food chain. Instead, alternative non-food sources are being searched and analyzed (Li and Lu 2021). On the other hand, the metabolites of autotrophic energy-harvesting organisms are also considered as a potential source of biofuels. However, their production is very limited. The cyano-bacteria (*Aphanizomenon, Arthrospira*, etc.) along with microalgae (viz. *Chlorella, Chlamydomonas*, etc.) and macroalgae (viz. *Porphyradentate, Laminaria*, etc.) are the potential source for next-generation biofuel (Singh and Gonzales-Calienes 2021). In recent years, algal lipids are being targeted for biofuel production. Algae vary in their lipid content and possess C_{14}-C_{18} carbon chain length. They may produce long-chain unsaturated fatty acids.

Further, triacylglycerols (TAGs) were found to be accumulated in many algae under stress conditions. The site of accumulation of fatty acids and TAGs are chloroplast and endoplasmic reticulum, respec-tively. These metabolites can be explored as biofuel sources. However, to compete with the other options available in the biofuel market, they must be produced in a sufficient amount under a short period along with the minimum cost.

5.3 Biodiesel

It is chemically made up of fatty acid methyl esters and can be produced from various biomass sources. Due to its beneficial aspects over oilseed crops, they are preferred over them. The microalgae species such as *Ankistrodesmus falcatus, A. fusiformis, Chlamydocapsa bacillus*, and *Kirchneriella lunaris* are preferred for their production (Nascimento et al. 2013). The cell wall of unicellular microalgae is dis-tinctly different from the higher animal and plants in respect to their lipids and fatty acids in their cell wall.

5.4 Biochemical Conversion and Thermochemical Conversion

Through the methods of biochemical conversion and thermochemical conversion, biofuels can be acquired from various sources like oil crops, the cooking oils, animal fats, and algal biomass. Microalgae are rich in the most important macronutrients, viz. nitrogen (N), Phosphorus (P), and potassium (K). These micronutrients contribute towards the accumulation of secondary metabolites that can be converted into biofuels (Kumar and Mukund 2018). The valuable strain of algae that are rich in carbohydrates, protein, lipids, and biomass and also similar in properties to that of fossil fuel can be used to produce an abun-dant amount of carbon-free important green bioenergies such as biodiesel, bioethanol, bio-methane, bio-hydrogen, and bio-butanol. The main benefits of microalgal production is it is biodegradable and uses CO_2 for photosynthesis, thereby helping in reducing environmental pollution as compared to fossil fuel, which produces harmful gas emissions such as sulphur oxide. Algal biomass with the aid of tech-nologies such as thermochemical, biochemical, and physicochemical can be used to produce biofuels. The thermochemical technologies can be further divided into three basic types of mechanism pyrolysis, gasification, and combustion. Thermochemical conversion technologies (TCCTs) are carried out in the absence of oxygen at temperatures of 300°C–650°C are used to change biomass into biodiesel in the form of solid or liquid or gases.

5.4.1 Transesterification

The conversion of algal oil to biodiesel is achieved by the transesterification process. Different alcoholic solvents, viz. amyl alcohol, methanol, butanol, ethanol, etc., are utilized in this process.

Ethanol and methanol are preferred commercially for this process (Gainey and Lawler 2021). Because of the high density of glycerol than biodiesel, it is required to be frequently or constantly removed from the reactor to make the equilibrium reaction. Using acid-alkali-catalyzed transesterification reaction results in 4,000 times faster conversion of triglycerides to methyl esters compared to the acid-catalyzed reaction.

5.4.2 Microalgae for Bioethanol Production

Many works are involved in research to find out the microalgae involved in the creation of biodiesel and having a higher content of carbohydrates as reserve polymers. Biofuels are procured from microalgae due to their exclusive potential. Algae are known for low content of lignin and hemicelluloses and have been assessed more appropriately for the creation of bioethanol. In recent times, more emphasis is given to making algae as the feedstocks by the process of fermentation for the production of bioethanol to make it as a usual crop such as corn and soybean. Their biofuels are eco-friendly, safe, and cost-effective. Moreover, they have high carbon dioxide sequestering efficiency and reduced GHG emissions (Tiwari and Kiran 2018).

Several marine algae are used to produces biogas through anaerobic transformation. By four key steps, this anaerobic conversion process is completed. Initially, obligate anaerobes, viz. *Clostridia* and *Streptococci*, release enzymes and help in hydrolyzing insoluble organic material of the higher molecular mass compounds. The second step is known as acidogenesis, in which acidogenic bacteria released some enzymes which are translated alcohols and volatile fatty acids (VFAs) of soluble organics biomass that finally get converted into acetic acid and hydrogen and further to methane and carbon dioxide by the methanogens.

5.4.3 Biohydrogen Production

Recently, for the generation of electricity and gaseous fuels, algal biohydrogen creation has been measured to be a general product to be used for the generation. Biohydrogen can be created by diverse processes like biophotolysis and photofermentation. Several photosynthetic microalgae, viz. *Chlorella*, *Botryococcus*, *Synechocystis*, *Anabaena*, *Nostoc*, *Tetraspora*, etc., are potential candidates used for its production.

Microalgae are ancient plants known to be the oldest life forms on earth. They are devoid of roots, stems, and leaves and have chlorophyll as their chief photosynthetic pigment. Microalgae are usually photosynthetic organisms that primarily use water, carbon dioxide, and sunlight to produce biomass and oxygen. Microalgae biomass is rich in carbohydrate and has been profoundly used as raw material for the production of different biofuels such as biodiesel, bioethanol, biohydrogen, and biogas in an economically valuable and environmentally friendly way. Microalgae belong to a varied assembly of prokaryotic and eukaryotic photosynthetic microorganisms; they are usually originated in habitats of marine and freshwater. They can be categorized as prokaryotic microalgae (Cyanobacteria), eukaryotic microalgae (green algae Chlorophyta), red algae (Rhodophyta), and diatoms (Bacillariophyta), which are proficient of growing quickly due to their simple structure and lesser nutrient requirements. Further, hydrogen gas is an extremely resourceful, proficient, and sustainable clean energy that may be used to replace the raw material due to its elevated energy yield when compared to traditional hydrocarbon fuels. Unit weight 66 of hydrogen gas can generate a heating value of 141.65 MJ/kg (Perry 1963). A select group of photosynthetic microorganisms has emerged with the capability to use light energy to drive hydrogen gas production from water.

5.4.4 Biogas Production

Biogas is produced by decomposing the biomass, viz. manure, plant waste, agricultural waste, etc., anaerobically. Nowadays, it is being popularized among the farmers due to its cost-effectivity and efficiency. Methane is the main component of biogas followed by CO_2 and other trace gases. Methanogens produce this gas through the methanogenesis process under anaerobic conditions. However, other associated processes are acetogenesis and acidogenesis.

Anaerobic digestion from algae in the creation of biogas has received notable notice due to the presence of various polysaccharides, viz. carrageenan, mannitol, agar, laminarin, etc. Moreover, they lack lignin that helps in their easy decomposition. Several species, viz. *Ulva*, *Euglena*, *Spirulina*, etc., are considered excellent for biogas production (Zhong et al. 2012; Saqib et al. 2013).

The hydrolysis step involves the conversion of larger biomolecules, viz. carbohydrates, lipids, proteins, and nucleic acids, into their smaller components. During acidogenesis, the monomers produced through hydrolysis are further dissociated. It forms volatile fatty acids that are degraded in the next step into acetic acid and other by-products. In the end, methanogens convert them into methane. There has been increasing attention paid to energy production from microalgae biomass. Microalgal biomass contributes to higher photosynthetic efficiency, higher biomass yield, and higher CO_2 fixation than higher plants. Sangeetha et al. (2011) reported the anaerobic digestion of green alga *Chaetomorpha litorea* with the generation of 80.5L of bio-gas/kg of dry biomass under 299 psi pressure. Due to the proteins present in algal cells resulting in ammonium production with little carbon to nitrogen ratio, they inhibit the growth of anaerobic microorganisms affecting biogas production. Further, sodium ions are also responsible for the inhibition of anaerobic microorganisms. Therefore, it is recommended that for anaerobic digestion of algal biomass, salt-tolerating microorganisms be used. Pre-treatment methods can be categorized as mechanical, thermal, chemical, biological, or a combination of these methods.

5.4.5 Mechanical Pre-treatment

The crystalline structure is broken into smaller sizes with the help of mechanical pre-treatment methods. As a result, the hydrolysis rate by enzyme degradation is increased. Mechanical methods often applied are ultrasonication and microwave application. Microwave application works on the electromagnetic energy between 300 MHz and 300 GHz; the oscillations of electric fields cause induction of heat because of friction generated causing the water to boil. The application of microwaves to this biomass causes dielectric polarization that results in changes in the hydrogen bonds by mechanical pre-treatment, which in turn affects the microalgae proteins. The cell disruption by ultrasound technique is most effective for microalgae pre-treatment. To create a cavity inside the cells, acoustic waves are used. Cell disruption is caused by the application of shock waves at high velocity and pressure that results in intense mixing. The efficacy of this pre-treatment is enhanced with frequency, power, and duration of time.

5.4.6 Thermal Pre-treatment

Thermal pre-treatment is a very commonly used method as it affects weak hydrogen bonds when mild temperatures (50°C–100°C) are employed; on the other hand, on increasing the temperature (>150°C) some complex polymers get dissolved. However, accumulation of stable compounds and higher energy consumption are the main disadvantages of this method. *Chlorella vulgaris* reportedly enhanced the solubility of protein and carbohydrate at 180°C but reduced the yield of methane and anaerobic biodegradation because of production of recalcitrant compounds.

5.4.7 Chemical Pre-treatment

Alkali and acid chemical pre-treatment methods are used for the polymerization and solubilization of cells. Specific biochemicals such as astaxanthin and c-phycocyanin are extracted using solvents. Alkali pre-treatment has been considered to increase the anaerobic biodegradation of microalgae. It also increases the exposed area of cellulose, which is caused due to its swelling, and decreases its crystallinity due to breakage of the glycosidic bond. Thus, alkali pre-treatment is best for microalgae with lesser lignin content. However, the cell wall of microalgae, carbohydrates, and proteins are hydrolyzed by acid pre-treatment. The lowered anaerobic biodegradation requires a high amount of energy, changes in pH, production of recalcitrant compounds, equipment corrosion, and expensive chemicals, which are the major issues with chemical pre-treatment methods.

5.4.8 Biological Pre-treatment

In this method, application of microorganisms and enzymes isolated from them are utilized in the bio-degradation of complex compounds. Enzymes like cellulase, proteases, lipases, esterases, and pectinase cause cellular hydrolysis of microalgae. But the chief drawback of this technique is it is not cost-effective and the procedure for enzyme formulation is complex of the optimum mixture.

5.4.9 Bio-Oil and Syngas Production

Thermochemical and biochemical methods are employed for the change of microalgal biomass into bio-oil. The kind and amount of biomass, economic consideration, and the type of energy required, and the reactions involved and its products are the major concerns on which the conversion of microalgal biomass into bio-oil depends. Production of bio-oil in the liquid phase takes place from microalgae in the anaerobic environment at very high temperatures. The chemical ingredients of bio-oil vary according to different microalgal biomass and methodology used, and the process is known as pyrolysis.

5.4.10 Hydrothermal Liquefaction (HTL)

Changes in the moistened microalgal biomass through hydrothermal liquefaction is an emerging technique to dry the algal biomass. HTL has higher energy recovery efficiency and lowers energy consumption rates from processing microalgal feedstocks to gasification and pyrolysis. HTL, without any removal of water content from microalgal biomass, which serves as the major source of energy input for biofuel generation. The technique produces four types of products, which are gaseous, aqueous, crude bio-oil, and crystalline. HTL is a widely accepted and practiced method for the conversion of microalgal feedstocks into bio-oil as it converts moistened microalgae into crude bio-oil avoiding excess consumption of energy. Moreover, the crude yield from this technique is greater as compared to other technologies employed (e.g., pyrolysis), resulting in hydrolysis of lipids, proteins, and carbohydrates in algae. Even the nutrients like Nitrogen, Phosphorus, and Potassium in algae are converted into salts or acids, which can further be reused for algal growth. Despite all this, the crude produced from algae feedstocks are disadvantageous, the water content in them is higher, they are highly viscous, have higher heteroatom content, and are not easily upgraded.

5.4.11 Gasification

It is the process to convert biomass into biogas through incomplete combustion. It involves the partial combustion of algal biomass to syngas at high temperatures. Syngas is a mixture of CO_2, CO, hydrocarbons, and hydrogen. Several catalysts, viz. nickel, can be used to increase the efficiency of the gasification process.

5.4.12 Pyrolysis

It is a method of thermal degradation under an inert environment. It produces gaseous and vapor phases along with crystalline char (Xing et al. 2021). This method is frequently used to convert biomass into the oil. Bio-oils from *Chlorella protothecoides* (55.3 wt%) and *Cladophora fracta* (48.2 wt%) were extracted by Demirbas (2011) using pyrolysis. Pyrolysis of *Spirulina* biomass can produce bio-oil (Chaiwong et al. 2013). The temperature requirement for this process lies in between 400°C and700°C.

5.5 Conclusion

Biofuels are renewable, biodegradable, and eco-friendly. Microalgae feedstocks have various advantages such as the fact that they are fast-growing, require a lesser land area for their growth and mass production, and possess enhanced lipids. Thus it can be used as an alternative for fossil fuel, thus meeting the never-ending energy demands all over the world while reducing the consumption of those fuel sources that increases the greenhouse gases. Because of certain other multiple advantages like increased oil content with greater production, it in future will replace all the other sources of biofuel and energy. Although there are many technologies available that can convert microalgae feedstocks into biofuels, the production of bio-oils by pyrolysis or liquefaction is very hopeful because of the less cost or resources needed. The generation of Biofuel can be made better by designing and developing techniques that aid in the better yield of the bio-oil.

REFERENCES

Azadbakht, M., A.S. Safieddin and M. Rahmani. 2021. Potential for the production of biofuels from agricultural waste, livestock, and slaughterhouse waste in Golestan province, Iran. *Biomass Conv Bioref.* https://doi.org/10.1007/s13399-021-01308-0.

Chaiwong, K., T. Kiatsiriroat, N. Vorayos and C. Thararax. 2013. Study of bio-oil and bio-char production from algae by slow pyrolysis. *Biomass Bioenergy* 56:600–606.

Demirbas, A. 2011. Competitive liquid biofuels from biomass. *Applied Energy* 88:17–28.

Gainey, B. and B. Lawler. 2021. The role of alcohol biofuels in advanced combustion: An analysis. *Fuel* 283:118915.

Hajjari, M., M. Tabatabaei, M. Aghbashlo and H. Ghanavati. 2017. A review on the prospects of sustainable biodiesel production: A global scenario with an emphasis on waste-oil biodiesel utilization. *Renewable and Sustainable Energy Reviews* 72:445–64.

Kumar N, N. Jeena, and H. Singh. 2019. Elevated temperature modulates rice pollen structure: A study from foothill Himalayan Agro-ecosystem in India. *3Biotech* 9:175.

Kumar, N., N. Kumar, A. Shukla, S.C. Shankhdhar and D. Shankhdhar. 2015. Impact of terminal heat stress on pollen viability and yield attributes of rice (*Oryza sativa* L.). *Cereal Research Communications* 43(4):616–626.

Kumar, R.R. and S. Mukund. 2018. Comprehensive review on lipid extraction methods for the production microalgae biofuel. *Journal of Algal Biomass Utilization* 9(1):48–61.

Li, Y.C. and C. Lu. 2021. Development perspectives of promising lignocellulose feedstocks for production of advanced generation biofuels: A review. *Renewable and Sustainable Energy Reviews* 136: 110445.

Nascimento, I.A., S.S.I. Marques, I.T.D Cabanelas, S.A. Pereira, J.I. Druzian, C.O. de Souza, et al. 2013. Screening microalgae strains for biodiesel production: lipid productivity and estimation of fuel quality based on fatty acids profiles as selective criteria. *Bioenergy Research* 6:1–13.

Perry, H. 1963. *Chemical Engineers' Handbook*. New York: McGraw-Hill.

Sangeetha, P., S. Babu and R. Rangasamy. 2011. Potential of green alga *Chaeto- morphalitorea*(Harvey) for biogas production. *International Journal of Current Science* 1:24–29.

Saqib, A., M.R. Tabbssum, U. Rashid, M. Ibrahim, S.S. Gill and M.A. Mehmood. 2013. Marine macroalgae *Ulva*: a potential feed-stock for bioethanol and biogas production. *Asian Journal of Agriculture and Biology* 1:155–163.

Singh D. and G. Gonzales-Calienes. 2021. Liquid Biofuels from Algae. In *Algae*, S.K. Mandotra, A.K. Upadhyay, and A.S. Ahluwalia (eds.), 243–279. Singapore: Springer, Press.

Tiwari, A., and T. Kiran. 2018. Biofuels from microalgae. In *Advances in Biofuels and Bioenergy*, N.M. Rao, and J. Soneji (eds.), 239–249. London: Intech Open Press.

Xing, X., K. Della and I.A. Udugama. 2021. Economic performance of small-scale fast pyrolysis process of coproducing liquid smoke food flavoring and biofuels. *ACS Sustainable Chemical Engineering* 9(4): 1911–1919.

Zhong, W., Z. Zhang, Y. Luo, W. Qiao, M. Xiao and M. Zhang. 2012. Biogas productivitybyco-digesting Taihu blue algae with corn straw as an external carbon source. *Bioresource Technology* 114:181–186.

6

New Insight in Biogas Plants:
A Sustainable Development Approach

Ratnakiran D. Wankhade

CONTENTS

6.1 Introduction

In developing countries like India, energy plays a major role in its overall development. There is a large increase in the demand for energy, which is mostly fulfilled by conventional energy sources such as natural gas, crude oil, and coal, but due to the depleting level of these energy sources, it will be challenging to fulfil these demands. The worldwide primary energy consumption till 2014 was up to 0.9 exajoules, or EJ (BP Statistical Review of World Energy, 2016). Biogas is the gas produced from the anaerobic digestion of organic material like animal waste, food waste, agriculture, and forest waste biomass. It is primarily used for cooking but also has use in electricity and vehicle use. It is targeted by the ministry of New and Renewable Energy to produce 48.55 MW energy till the year 2022 from biogas plants (BP Statistical Review of World Energy, 2016).

There is a large potential in India and the world for employing anaerobic digestion for biogas production from waste treatment. Anaerobic digestion mainly consists of the various biological processes in which the biodegradable material is broken down with the help of microorganisms in the absence of oxygen. The primary biochemical processes that take place are hydrolysis, acidogenesis, and methanogenesis. Researchers have observed the effect of the different feedstock materials. The farmyard manure as feedstock material shows 2–5 times less methane production than the switchgrass, corn stover, and wheat straw. Methane yield was 15 and 17 L/kg for waste paper and pine, respectively (Brown et al., 2012; Singh et al., 2020; Kumar et al., 2021). Biogas is an end product obtained from the anaerobic digestion of hydrocarbons, which are mainly comprised of methane (CH_4), carbon dioxide (CO_2), and some other components (Walsh et al., 1988). It can be defined in a more straightforward form as a result of the complete microbial degradation of organic matter under anaerobic conditions. This organic matter may be waste from different sources like landfills, industry, or sewage sludge. Also, apart from non-fossil fuels energy production, there are many reasons for utilizing anaerobic digestion. Generally, organic waste is naturally degraded by micro-organism in landfills, mainly producing methane and carbon dioxide. Methane gas is a very hazardous pollutant, having 23 times greater greenhouse effect than carbon dioxide (Petersson and Wellinger, 2009a). Hence, it is far better to collect the biomethane for combustion, which will reduce environmental pollution, and the rest of the product can be further used as fertilizer. Biogas also contains some other gases like siloxanes, hydrogen sulphide, ammonia, halogenated hydrocarbon, and methane and carbon dioxide. The impurities can cause health and environmental issues. Also, they can damage processes, equipment pipelines, and nozzles. These impurities will affect the final CH_4% in biogas, which results in a lower calorific value. Hence, improving the quality of the biogas from an economic and environmental point of view is very necessary. The upgradation is carried out by pre-treatment, which similar to the pre-treatment of fossil fuels before utilization.

6.2 Biogas Production

The various steps of conversion of organic matter into the biogas by microorganism are through the series of metabolic stages as hydrolysis, acidogenesis, acetogenesis, and methanogenesis (Figure 6.1). In hydrolysis, complex organic compounds such as lipids, proteins, and polysaccharides are converted into soluble monomers or oligomers like amino acids glycerol, long-chain fatty acid, sugars, etc., also known as liquefaction. In this process, the hydrolytic or fermentative bacteria release extracellular enzymes. The acidogenic bacteria ferment the simple soluble compounds into a mixture of carbon dioxide (CO_2), hydrogen (H_2), alcohol, and low molecular weight volatile fatty acids (VFAs) like propionic and butyric acids. This process is known as acidogenesis. In this process, alcohols and VFAs are anaerobically oxidized by hydrogen-producing acetogenic bacteria into H_2, CO_2, and acetate. Hydrogen-oxidizing acetogenic bacteria known as homoacetogens can also form acetate by H_2 and CO_2 (Khanal, 2008). The acetate, H_2, and CO_2 are transformed into a mixture of CH_4 and CO_2 by acetotrophic and hydrogenotrophic methanogens in the final stage. Acetate is utilized by autotrophic methanogens as a substrate in a process known as acetotrophic methanogenesis. In the hydrogenotrophic methanogenesis process, the hydrogenotrophic methanogens reduce CO_2 by using H_2 as an

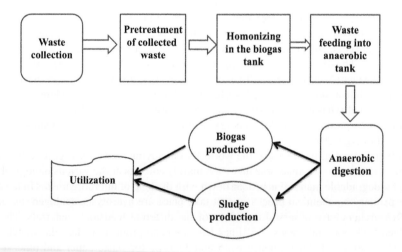

FIGURE 6.1   Biogas production process

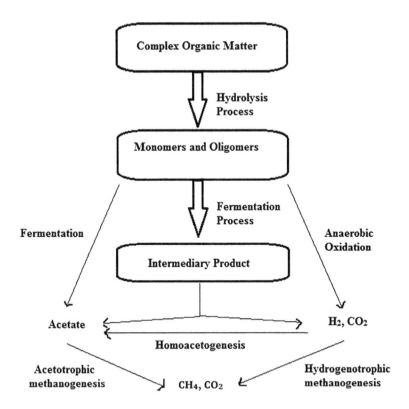

FIGURE 6.2 Biomethanation process.

electron donor. The decarboxylation of acetate produced 70% of the total CH_4 produced, while the remaining is produced by CO_2 reduction (Gerardi, 2003; Zieminski and Frac, 2012). Some amount of CH_4 is produced from butyric, propionic, and formic acid, and also from other organic substrates by methanogens (Figure 6.2) (Woodard, 2006).

6.3 Biogas Composition

The biogas composition depends mainly on the feedstock and the operating condition of the biogas digester. Generally, biogas contains CH_4 50%–75%, CO_2 25%–50%, along with other trace components like hydrogen sulphide (H_2S), water vapour (H_2O), and ammonia (NH_3). The properties and typical composition of the raw biogas are shown in Table 6.1. The only component in the biogas related to heat, which is the calorific value, is the CH_4.

6.4 Biogas Technology

The current organic matter decaying biogas plant was used to heat water, a practise that began thousands of years ago in Assyrian bath houses (Bond and Templeton, 2011). After the 1950s, interest and rigorous research on the technology emerged to the well-known digester designs like the Grama Laxmi III (proposed by Joshbai Patel in India, which later served as a currently popular floating-drum digester) (Akinbami et al. 2001). In countries like India and China, domestic digesters gained momentum. Also, Government support for the household digester installation is available. About 26.5 and 4 million household digesters in China and India, respectively, were there by the year 2007. The Chinese fixed dome

TABLE 6.1

Biogas Chemical Composition and Components Properties (Muylaert et al., 2000)

Components	Concentration (%)	Properties
Water Vapour	1–5	Facilitate corrosion in the presence of CO_2 and sulphur dioxide (SO_2)
N_2	0–5	Decreases heating value
NH_3	0–500	NOx Emissions during combustion
$H2_S$	0–5,000	Corrosive Sulphur dioxide emission during combustion
CO_2	25–50	Decrease heating value Corrosive, especially in the presence of moisture
CH_4	50–75	Energy Carrier

digester and the Indian floating drum digester are the two most common digester designs among all digester designs. Generally, these digesters are used for converting human and animal wastes of one household to biogas for cooking and lighting.

The average volume of the digester is approximately 5–7 m³ and provides about 0.5 m³ biogas per m³ digester volume (Omer and Fadalla, 2003). In a floating drum digester, the construction material is steel and concrete, while in the case of a fixed dome digester, bricks and stones are used for construction. Only the cover is above ground, and the rest of the components are housed below ground in a floating drum digester. As regards the working principle, both the digesters are quite similar. The feedstock is fed through the inlet pipe after mixing in a pit or directly in the digester. The biogas obtained over the slurry is collected and passed out through the pipe connected above the top of the digester. The slurry leaves the digester through the outlet pipe, which is collected in the displacement tank. The digester has one or two compartments depend on the type of digester where substrate is left for a retention time of 20–30 days. In a floating drum digester, floating steel cover maintains the pressure while the fixed dome gas is roughly kept at constant volume. In both these digester system, there is a lack of proper mixing and temperature control. Also, there is no provision for the removal of settled inert material. The simple construction and the lack of moving parts make this system easy to operate and maintain (Surendra et al., 2010). Hence, to minimize the cost, the effort was made by developed countries for designing low-cost polyethylene tubular digesters (Khan, 1996). There is also a drawback of lack of mixing and heating system in the polyethylene tubular digesters. In this system, to avoid the sophisticated monitoring needs, it is fabricated using readily available materials, generally plastic bags for the main tank and polyvinyl chloride (PVC) pipes for biogas collection. In this system, feedstock passes through a tubular polyethylene or PVC bag. At the same time, biogas is collected utilizing a gas pipe connected to the headspace, while biogas is collected by a gas pipe connected to the headspace. In the cold mountain area, for maintaining the higher processes temperature and minimizing overnight temperature fluctuation, the tubular plastic bag is often buried in a trench and covered with gable or roofed shed to provide crude insulation (Ferrer et al., 2011). The dimension and design criteria change as per the location. At the high altitudes with colder temperatures, there is a longer retention time of 60–90 days (Karagiannidis, 2012). Simple design, less manpower, and ease of installation make this design acceptable and affordable for household application in many developing countries (Poggio et al., 2009). The fragile nature of the polyethylene film makes the digester prone to damage (Luer, 2010).

6.5 Biomass Resources for Biogas Production

The raw material for biogas production is essential. Generally, human excreta, animal manure, sewage sludge, organic fraction of municipal solid waste (OFMSW), energy crops, and forest residue can be used as raw material. In the context of energy generation from these resources, it has been found to be more competitive than alcohol-based liquid biofuel (Chynoweth et al., 2018). Also, anaerobic digestion minimizes greenhouse gases, which provide the opportunity for organic waste remediation. Table 6.2

TABLE 6.2

Biogas Production Potential of Various Organic Sources

S. No.	Source	Specific Biogas Production (Nm³/Kg DM)
1.	Sheep Manure	0.30–0.40
2.	Cattle Manure	0.20–0.30
3.	Pigs Manure	0.20–0.50
4.	Chicken Manure	0.31
5.	Night soil	0.38
6.	Forage Leaves	0.50
7.	Algae	0.32
8.	Grass Ensilage	0.60–0.70
9.	Grass cutting From lawns	0.70–0.80
10.	Vegetable waste	0.40
11.	Slaughter House Waste	0.30–0.70
12.	Water Hyacinth	0.40
13.	Potato mash, Potato pulp, Potato peelings	0.30–0.90
14.	Maize Straw	0.40–1.00
15.	Rice Straw	0.55–0.62
16.	Maize ensilaged	0.60–0.70
17.	Hay	0.50
18.	The organic fraction of municipal solid waste	0.10–0.93 (Nm³/kg wet weight.)
19.	Waste from paper and carton production	0.20–0.30
20.	Leftovers (kitchen)	0.40–1.00
21.	Molasses	0.30–0.70
22.	Oilseed residuals(pressed)	0.90–1.00

shows the biomass production potential of different biomasses. It is also observed that the developed countries are already using a million tonnes of organic waste from agricultural processes and municipal and industrial waste for biogas production (Gunaseelan, 1997), yet, much of the volume of the organic waste remain unused in developing countries.

6.6 Biomass Upgradation

The biomass upgradation is very necessary to obtain it in its pure form and remove the impurities. The main upgradation techniques are adsorption, absorption, membrane separation, and cryogenic methods. These technologies are primarily used for CO_2 separation. Also, pre-upgradation is required to remove high contaminant like Siloxanes, H_2O, and H_2S.

6.6.1 Adsorption

In this process, the transfer of solute in the gas stream to the surface of an absorbent material due to physical or van der Waals forces is carried out. Some undesirable gases like CO_2 are separated from biogas in pressure swing adsorption (PSA) under elevated pressure using adsorbent material, after which the gases are desorbed by reducing the pressure (Cavenati et al., 2005). Cormac, Gasrec, Carbotech, Arona, Guild Associates, and Xebec, Inc. are well-known companies that develop and commercialize this technology at low and high capacity (flow rate of 10–10,000 m³/h of biogas) (Hullu et al., 2008).

6.6.2 Absorption

The main factor for absorption is the solubility of various gas components in the liquid solvent. In this system, raw biogas meets a counter flow of liquid in the column filled with packing material for increasing the contact area between gas and liquid. In this process, it is found that CO_2 is more soluble than CH_4, while the liquid that is left the column has an increased concentration of CO_2 (Cozma and Ghinea, 2013). High-pressure water scrubbing (HPWS) and organic physical scrubbing (OPS) fall under this type of physical absorption. On the other hand, amine scrubbing (AS) and inorganic solvent scrubbing (ISS) are under chemical absorption (Chen et al., 2018).

6.6.2.1 High-Pressure Water Scrubbing (HPWS)

It is the most common and well-established technology used for removing the CO_2 and H_2O gases from the biogas because these gases are more soluble in H_2O than CH_4. The system's operating pressure is 10 bar during the process (Cozma et al., 2014) at the bottom of the packed column where the biogas is fed, and the water is fed countercurrently. The physical absorption takes place as per Henry's law, which states that the amount of any dissolved gas is directly proportional to its partial pressure in the gas stream at a constant temperature (Petersson and Wellinger, 2009b). It is also helpful to remove the H_2S due to its increased solubility in water than CO_2 (Abatzoglou and Boivin, 2009).

6.6.2.2 Organic Physical Scrubbing (OPS)

The main difference of Organic physical scrubbing (OPS) from water scrubbing is that it uses the organic solvents like N-methyl pyrrolidone (NMP), polyethylene glycol ethers (PEG), and methanol (CH_3OH) to absorb CO_2 instead of water. CO_2 has a five-times higher solubility in PEG than water for the same upgradation capacity (Tock et al., 2010) due to which it requires less organic solvent and pumping requirement. In this process, CO_2, H_2O, H_2S, O_2, N_2, and other halogenated hydrocarbons are also removed, but the prior removal of H_2S is recommended (Zhao and Leonhardt, 2010). The biogas is compressed at a compression pressure of 6–8 bar and, after cooling, injected into the absorption column. The counter-current flow of gas and liquid is made by supplying the organic solvent to the top of the column, which is also cooled before injecting in to the column for maintaining the low temperature up to 20°C.

6.6.2.3 Chemical Scrubbing Process (CSP)

In this process, the most common amines used for removing acidic gases like CO_2 and H_2S are monoethanolamine (MEA), methyl diethanolamine (MDEA), and diethanolamine (DEA). This process is the reversible reaction between solvent and absorbed substance. Presently, a mixture of MDEA and piperazine (PZ), called activated MDEA (AMDEA) is commonly used in this process. The amino scrubber system consists of an absorber in which CO_2 is absorbed from the biogas and a stripper in which CO_2 is separated from waste amine solution by heating in reduced pressure (Kismurtono, 2011). In the absorber, biogas is fed from the bottom side, while amine solution is supplied to the top of the column to make the counter-current flow contact. After reaction of CO_2 with amine solution biogas get absorbed. This process is the exothermic process in which the absorber temperature ranges from 20°C–40°C to 45°C–65°C.

6.6.3 Membrane Separation (MS)

It the quite a popular method for the last 40 years and has become a large part of the market share (Makaruk et al., 2010). In this method, the membrane acts as a permeable barrier that allows specific compounds to pass through differently and is based on applied driving forces such as pressure, temperature, electric charges of different species, and difference in concentration to control the permeability.

Typically, there are two different models: solution-diffusion and pore-flow model (Jiang and Chung, 2006). The permeate dissolves into the membrane material and diffuses through the because of concentration difference. After that, permeates are separated by pressure-driven convective flow through small pores. In the transportation of polymeric membranes, the solution diffusion model is frequently used (Kentish, 2008).

In the case of biogas upgradation, CO_2 permeates through the membrane while CH_4 remains on the inlet side as retentate. It is observed that the membrane gas separation process is more beneficial when

CO_2 content is high and gas flow is low (Baker and Lokhandwala, 2008). The main reason for this method's popularity is the cheap process, including low capital cost, simple and compact membrane equipment setup, and less energy demand. There are three different types of membrane used for biogas purification, i.e., polymeric, inorganic, and mixed matrix membranes (MMMs).

6.6.4 Cryogenic Separation (CS)

The working principle of the cryogenic separation is the pressure and temperature difference. Gases like CO_2 and H_2S liquefy under different pressure and temperature conditions. The cryogenic separation of biogas carried out under very low temperatures, i.e., 170°C, and high pressures, up to 80 bar. The boiling point of the CH_4 is −161.5°C at 1 atm, while it is −78.2°C for CO_2, which allows the separation of both gases by liquefying (Andriani and Wresta, 2014). These operating conditions are maintained by a series of compressors and heat exchangers (Zanganeh et al., 2009). The main drawback of this process is the high capital and operational cost due to the use of different processes equipment with high energy requirements. There are generally four steps in this process of upgradation biogas to natural gas quality. In the first step, unwanted components are removed from the biogas, while in the second step, biogas is compressed to 1,000 kPa and subsequently cooled to −25°C. The third step includes cooling the biogas up to −55°C, and liquefied CO_2 is removed from the gas mixture. In the final step, the remaining gas stream is further cooled up to −85°C at which CO_2 reaches a solid form and is removed out, while the purified gas is depressurized and can be used for various applications (Ryckebosch et al., 2011).

6.7 Biogas Application

Biogas has various applications like heating, transportation fuel, combined heat and power generation, or upgraded to natural gas quality for diverse application. In many countries having household-scale digesters, the end-use of biogas is limited to lighting and cooking. Biogas produced from large-scale institutional plants in many developing countries is being used for electricity generation through fuel cells or CHP engines. In India, a community biogas digester is used to power a modified diesel engine and run an electrical generator. The methane (in biogas) burns with a clean blue flame that is much hotter than fire fuelled by traditional resources; hence, adopting biogas is increasing day by day. Mostly in rural areas, biogas cookstoves are being used successfully. In Nepal, it has been estimated that 0.33 m³ of biogas is required to fulfil the cooking demand per capita per day (Bauer et al., 2013). The second most common use of biogas is for lighting, especially in regions which do not have an electrical grid connection.

6.8 Conclusion and Future Recommendation

The biogas plant is available in different ranges and types, so it is imperative to select the biogas plant as per need and requirement. There are various biomass resources for biogas generation, but researchers can still pay attention to finding out new resources for more efficient biogas production. It is also of critical consideration to know more on biogas upgradation, storage, and use. The application of biogas is increasing in various sectors day by day; however, still spreading awareness regarding the biogas utilization is necessary as it is a clean, environment-friendly and renewable source of energy. Also, there is a need to fill the knowledge gap between the new biogas generation method and the users. In the future, biogas can play a vital role in fulfilling everyday energy demand.

REFERENCES

Abatzoglou, N. and S. Boivin. 2009. A review of biogas purification processes. *Biofuels, Bio prod Biorefining* 3(1):42–71.

Akinbami, J.F.K, M.O. Ilori, T.O. Oyebisi, I.O. Akinwumi and O. Adeoti. 2001. Biogasenergy use in Nigeria: currentstatus, future prospects and policy implications. *Renewable and Sustainable Energy Reviews* 5:97–112.

Andriani, D. and A. Wresta. 2014. A review on optimization production and upgrading biogas through CO2 removal using various techniques. *Applied Biochemistry and Biotechnology* 172(4):1909–28.

Baker, R. and Lokhandwala, K. 2008. Natural gas processing with membranes: an overview. *Industrial & Engineering Chemistry Research* 47(7):2109–21.

Bauer, F., C. Hulteberg, T. Persson and D. Tamm. 2013. Biogas upgrading – review of com- mercial technologies. *SGC Rapp* 270:1–83.

Bond, T. and M.R. Templeton. 2011. History and future of domestic biogas plants in the developing world. *Energy for Sustainable Development* 15:347–54.

BP Statistical Review of World Energy (bp.com/statistical review) June 2015 (Accessed 3.05.2016).

Brown, D., J. Shi, and Y. Li. 2012. Comparison of solid-state to liquid anaerobic digestion of lignocellulosic feedstock for biogas production. *Bioresource Technology* 124:379–86.

Cavenati, S., C. Grande and A. Rodrigues. 2005. Upgrade of methane from landfill gas by pressure swing adsorption. *Energy Fuels* 19(6):2545–55.

Chen, X., H. VinhThang, A.A. Ramirez, D. Rodrigue and S. Kaliaguine. 2018. Membrane gas separation technologies for biogas upgrading. *RSC Advances* 5(31):24399–448. http://dx.doi.org/10.1039/C5RA00666J.

Chynoweth, D.P., J.M. Owens and R. Legrand. 2018. Renewable methane from anaerobic digestion of biomass. *Renew Energy* 22:1–8.

Cozma, P. and C. Ghinea. 2013. Environmental impact assessment of high pressure water scrubbing biogas upgrading technology. *CLEAN-Soil, Air Water* 41(9):917–27.

Cozma, P., W. Wukovi, A. Friedl and M. Gavrilescu, 2014. Modeling and simulation of high pressure water scrubbing technology applied for biogas upgrading. *Clean Technologies and Environmental Policy* 17:373–91 http://dx.doi.org/10.1007/s10098-014-0787-7.

Ferrer, I., M. Garfí, E. Uggetti, L. Ferrer-Martí, A. Calderon, and E. Velo, 2011. Biogas production in low-cost household digesters at the Peruvian Andes. *Biomass Bioenergy* 35:1668–74.

Gerardi, M.H. 2003. *The Microbiology of Anaerobic Digesters.* New Jersey, United States: John Wiley and Sons Inc.

Gunaseelan, V.N. 1997. Anaerobic digestion of biomass for methane production: A review. *Biomass Bioenergy* 13:83–114.

Hullu, J., J. Waassen and P. Van Meel. 2008. Comparing different biogas upgrading techniques. *Eindhoven University of Technology* 56:56–100.

Jiang, L.Y. and T.S. Chung. 2006. Kulprathipanja S. An investigation to revitalize the separation performance of hollow fibers with a thin miXedmatriX composite skin for gas separation. *Journal of Membrane Science* 276:113–25. http://dx.doi.org/10.1016/j.memsci.2005.09.041.

Karagiannidis, A. 2012. *Waste to Energy: Opportunities and Challenges for Developing and Transition Economies.* London: Springer.

Kentish, S. 2008. Carbon dioXide separation through polymeric membrane systems for flue gas applications. *Recent Patents on Chemical Engineering* 1(1):52–66.

Khan, S.R. 1996. Low cost biodigesters. Programme for research on poverty alleviation, Grameen trust report.

Khanal, S.K. 2008. *Anaerobic Biotechnology for Bioenergy Production: Principles and Applications.* Ames, Iowa: John Wiley & Sons, Inc.

Kismurtono, M. 2011. Upgrade biogas purification in packed column with chemical ab- sorption of CO_2 for energy alternative of small industry (UKM-Tahu). *International Journal of Engineering & Technology* 11:59–62.

Kumar, A., P. Kumar, H. Singh, S. Bisht and N. Kumar. 2021. Relationship of physiological plant functional traits with soil carbon stock in temperate forest of Garhwal Himalaya. *Current Science* 120(8):1368–1373.

Luer, M., editor. 2010. Installation manual for low-cost polyethylene tube digesters.GTZ/EnDev, Germany.

Makaruk, A., M. Miltner and M. Harasek. 2010. Membrane biogas upgrading processes for theproduction of natural gas substitute. *Separation and Purification Technology* 74:83–92. http://dx. doi.org/10.1016/j. seppur.2010.05.010.

Muylaert, M.S., J. Sala and M.A.V. Freitas. 2000. Consumo de energia e aquecimento do planeta: Análise do mecanismo de desenvolvimentolimpo (MDL) do Proto- colo de Quioto. Case studies. Rio de Janeiro: Post-graduate Engineering Programs Coordination (COPPE).

Omer, A.M. and Y. Fadalla. 2003. Biogas energy technology in Sudan. *Renew Energy* 28:499–507.

Petersson, A. and A. Wellinger, 2009a. Biogas upgrading technologies developments and innovations. *IEA Bioenergy* 37;1–15.

Petersson, A. and A. Wellinger. 2009b. Biogas upgrading technologies-developments and innovations. In Task 37, Energy from biogas and landfill gas. IEA Bioenergy :1–19.

Poggio, D., I. Ferrer, L. Batet and E. Velo, 2009. Adaptacin de biodigestorestubulares de plstico a climasfros. Livest Res Rural Develop, 21 (Article 152) Available from: http://www.lrrd.org/lrrd21/9/pogg21152.htm [retrieved 11.10.11].

Ryckebosch, E., M. Drouillon and H. Vervaeren. 2011. Techniques for transformation of biogas to biomethane. *Biomass Bioenergy* 35(5):1633–45. http://dx.doi.org/10.1016/j.biombioe.2011.02.033.

Singh, H., M. Yadav, N. Kumar, A. Kumar and M. Kumar 2020. Assessing adaptation and mitigation potential of roadside trees under the influence of vehicular emissions: A case study of Grevillea robusta and Mangifera indica planted in an urban city of India. *PLoS ONE* 15(1): e0227380.

Surendra, K.C., S.K. Khanal, P. Shrestha, and B. Lamsal. 2010. Current status of renewable energy in Nepal: Opportunities and challenges. *Renewable & Sustainable Energy Reviews* 15:4107–17.

Tock, L., M. Gassner and F. Maréchal. 2010. Thermochemical production of liquid fuels from biomass: Thermo-economic modeling, process design and process integration analysis. *Biomass Bioenergy* 34:1838–54.

Walsh, J.L., M.S. Smith, C.C. Ross, and S.R. Harper. 1988. *Biogas Utilization Handbook.* Georgia Tech Research Institute Publisher Atlanta, Georgia, p. 133.

Woodard, F. 2006. *Industrial Waste Treatment Handbook.* 2nd ed. Oxford: Butterworth-Heinemann.

Zanganeh, K.E., A. Shafeen and C. Salvador. 2009. CO_2 capture and development of an advanced pilot-scale cryogenic separation and compression unit. *Energy Procedia* 1:247–52. http://dx.doi.org/10.1016/j.egypro.2009.01.035.

Zhao, Q. and E. Leonhardt. 2010. Purification technologies for biogas generated by anaerobic digestion. *Center for Sustaining Agriculture & Natural Resources* 1:21–38.

Zieminski, K. and M. Frac. 2012. Methane fermentation process as anaerobic digestion of biomass: Transformations, stages and microorganisms. *African Journal of Biotechnology* 11:4127–39.

7

Efficient Utilization of Biomass Energy for Mitigating Future Climate Change

Aditee Das and Roop Jyoti Das

CONTENTS

7.1 Introduction

Climate change is one of the critical challenges facing the world in the 21st century. It engages the energy sector in a close-knit because energy is central to the problem and its resolution. Energy-related emissions account for over two-thirds of anthropogenic greenhouse gas (GHG) emissions and over 80% of worldwide emissions of CO_2 as a direct result of fossil fuel combustion (Kaygusuz, 2010). The different types of energy sources will play an essential role in the world's future. The world's energy markets rely heavily on the fossil fuels such as coal, petroleum crude oil, and natural gas as energy sources. The formation of fossil fuel reserves of coal in the earth takes millions of years, but it is rapidly depleting as a result of human consumption. Biomass is the only naturally existing renewable carbon resource that has the potential for the replacement of fossil fuels. Essential biomass energy sources are wood and wood waste, municipal solid waste, animal waste, aquatic plants and algae, crops, and their waste by-products. Unlike fossil fuel deposits, biomass is a renewable energy source that is likely to gain momentum in the future (Klass, 2004). Since the signing of the United Nations Framework Convention on Climate Change (UNFCCC) in Rio de Janeiro in 1992, the interest in solar-based renewable energy sources has intensified. Among these energy sources, energy from biomass is considered one of the potential energy sources to replace some fossil fuels whose combustion is the primary source of anthropogenic greenhouse gases, specifically CO_2 (Bilgen, Keles, and Kaygusuz, 2007). According

DOI: 10.1201/9781003175926-7

to IEA (2017), presently biomass is the largest renewable energy source used by mankind. We should not underestimate biomass's potential as its utilization has gained momentum in recent years due to the continuous exploitation of conventional fossil fuels. Biomass has often been looked upon as a promising renewable energy source to mitigate climate change. However, it is unclear what effect biomass energy will have on global surface temperatures if it replaces fossil fuels. As a result, this chapter will shed some light on whether biomass energy will be able to address future climate change through efficient utilization.

7.2 Biomass Energy: An Overview

Biomass energy is considered one of humanity's earliest sources of energy. It is the very first form of energy other than food that the humankind learned to use (Sürmen, 2003).The use of fire for energy sources from biomass has been in use for 10,000 years. The product of biomass energy is bioenergy and biofuels. Bioenergy refers to solid biomass used primarily for industrial applications (power, heat), domestic uses (cooking, heating), and small- and large-scale use. Biofuels refer to liquid biofuels (bioethanol and biodiesel) generally used in road transport. (Rosillo-Calle, 2016). The energy from fossil fuels and bioenergy came originally from solar energy. The main difference between the two forms of energy is that fossil fuels are extracted from underground excavation, i.e., from mining.

In contrast, biomass energy is extracted from living or dead plant materials. If properly managed, it has the potential to surpass fossil fuel consumption shortly. Biomass includes 'phytomass' or plant biomass and 'zoomass' or animal biomass. When taken up by plants and converted by photosynthesis into chemical energy, Sun's energy is 'fixed' or stored in the form of terrestrial and aquatic vegetation. The vegetation, when grazed by animals, gets converted into animal biomass and excreta. The excreta of terrestrial animals, traditionally dairy animals, can be used as a source of energy. In contrast, aquatic animal's excreta do not contribute to energy production as it gets dispersed and challenging to collect. Generally, animal biomass contributes very little to the CO_2 emission (Abbasi and Abbasi, 2010).

According to the National Institute of Statistics, the biomass is the organic matter that can become energy sources. They can be used directly (wood energy) or right after methanation of the organic matter (biogas) or (biofuel). The term biomass refers to all plants or, to be more specific, all materials derived from growing plants. Over the last decades, biologists and ecologists lost the cartel of defining biomass; therefore, engineers and industrialists adopted and extended it to include such diverse sources as algae, municipal solid waste, food wastes, and agro-industrial by-products. Such a breakthrough shift promoted by the valuable research and development facilitated the production of various commodities from biomass, especially biofuels, after the oil crisis in the 1970s (Sillanpää and Ncibi, 2017). Simply put, biomass is the dry weight of all organic material, living or dead, above or below the soil surface (The State of Canada's Forests, 2003). Biomass energy is the fuel or steam derived from the biomass's direct combustion to generate electricity, mechanical power, or industrial process heat. Biomass can be transformed into energy, biofuels, or burned to produce heat or electricity to cope with the energy crisis.

Additionally, biomass has various other uses such as food and feed, forestry products (pulpwood), and other industrial applications that are important to support the world's economy as a whole (Sürmen, 2003; Kumar et al., 2020, 2021). Biomass contains cellulose, hemicellulose, lignin, and small amounts of other organics and inorganics—the organic components in biomass are vital in developing processes for producing fuels and chemicals (Abbasi and Abbasi, 2010). The combination of cellulose, hemicelluloses, and lignin (polymers) is called 'lignocellulose'. Biomass energy (Figure 7.1) is derived from directly burning it or from one or another fuel. However, some species are proven scientifically as providing a better fuel quality at lesser costs than other species.

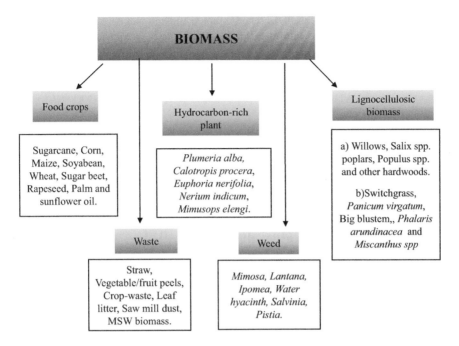

FIGURE 7.1 Sources of biomass for energy production.

7.3 Traditional vs. Modern Biomass Energy

In developing countries, low-efficiency traditional biomass is used for cooking, lighting, and heating by the underprivileged population. Traditional cooking and using low-efficiency stoves have low combustion efficiency. It gives rise to large amounts of incomplete combustion products (CO, methane, particle matter, and volatile organic compounds) that negatively impact climate change and air pollution. In addition to this, the biomass harvested renewably cannot be considered GHG neutral due to incomplete production products such as black carbon. There is estimation worldwide that traditional fuel combustion causes approximately 30% of the warming due to black carbon and carbon monoxide emissions from human interventions, about 15% of ozone-forming chemicals.

In contrast, methane and CO_2 contribution in emission is a few percent (Miltenburger, Borgstrom, and Eklund, 2018; Costello et al., 2009; Haines et al., 2009; Wilkinson et al., 2009). High-efficiency modern bioenergy uses solids, liquids, and gases as secondary energy sources to generate heat, electricity, combined heat and power (CHP), and fuel for transportation. Biomass-derived gases (methane from anaerobic digestion of agricultural residues and municipal solid waste) generate electricity, heat, or both. For global road transport, liquid biofuels such as ethanol and biodiesel are utilized (IPCC, 2011). Traditional biomass has fewer implications for future climate change in comparison to modern bioenergy. As modern bioenergy's efficiency is almost double that of conventional biomass, an in-depth analysis of modern bioenergy is done in the next section of the chapter.

7.4 Modern Bioenergy Technologies for Mitigating Future Climate Change

At present, there is enough scientific evidence which indicates that climate change is a serious issue of the 21st century. The earth's climate is changing rapidly, and the credit for the change goes to greenhouse gas emissions due to anthropogenic activities. According to NASA (2021), the present atmospheric CO_2

concentration is about 415 parts per million (ppm), which is almost double the concentration of the pre-industrial levels. As far as the commitments made under Paris Agreement, it is necessary to keep global warming well below 1.5°C–2°C, as compared to the pre-industrial level (UNFCCC, 2016). Every year there is an increase of about 2 ppm, and the data will soon exceed the average figures in the coming years (NOAA, 2020). With an unprecedented rise in global temperature, there are severe effects on environmental health. As far as the history of the earth is concerned, the increase in global CO_2 levels has never been a topic of debate. Climate change has severe implications on the economy and environment of the world. Proper implementation of robust policies for carbon-neutral and carbon-negative emissions to overcome the environmental and economic constraints is the need of the hour. One such technology that will align with the targeted goal of the Paris agreement to keep global warming well below 1.5°C–2°C is bioenergy with carbon capture storage (BECCS) (Hansen et al., 2020).

7.5 Bioenergy with Carbon Capture and Storage (BECCS)

Bio-Energy with Carbon Capture and Storage (BECCS) is a technology that incorporates biomass energy systems with geological carbon storage (IPCC, 2012; AR4-IPCC, 2007). R. H. William mentioned the first BECCS technology in scientific publications in the 1990s. Since then, the BECCS technology became a form of Carbon Capture and Storage (CCS) technology. BECCS is a potential technology to create permanent negative carbon emissions, i.e., removing CO_2 from the atmosphere (IPCC, 2011).

7.6 Carbon-Negative Energy with BECCS Technology

The one thing that makes BECCS unique is its potential to result in carbon-negative emission of CO_2. The biomass energy coupled with carbon capture can effectively remove carbon from the atmosphere (Read and Lermit, 2005).Bioenergy is a renewable source of energy, and it serves as a carbon sink during its growth. The biomass combusted or processed during industrial operations re-releases the CO_2 into the atmosphere. It results in a net-zero emission of CO_2 during the process (Cassman and Liska, 2007). The IPCC report on CCS technology estimated that more than 99% of carbon dioxide stored in geologic formation is estimated to be held for more than 1,000 years (Celia et al., 2005).

On the other hand, significant carbon sinks such as the ocean, soil, and trees may experience adverse climate change feedback at increased temperatures. BECCS technology is a promising candidate to provide a better solution by storing CO_2 in geological formations to overcome the uncertainties (Global Status of BECCS, 2010). Anthropogenic activities have released tremendous amounts of CO_2 to be absorbed by conventional sinks such as the ocean, soil, and tree to reach the Paris Agreement target. Even in the most ambitious low-emission scenarios, there will be significant additional emissions during the 21st century. In relevance to the present scenario, BECCS has been suggested as a potential technology to reverse the current emission trend and create a global carbon-negative emissions system (Lindfeldt and Westermark, 2008; Azar et al., 2006; Hare and Meinshausen, 2006; Obersteiner, 2001). This further implies that the emissions would not only be zero but negative. This technology aligns with the Paris targets and goes beyond it to cap down carbon dioxide levels. The carbon-negative emission scenario due to BECCS employment is shown in Figure 7.2a and b.

7.7 Challenges Faced by BECCS Technology

BECCS will require extensive land use for biomass production in order to be deployment globally. This implies that it will have severe constraints such as land and water scarcity, biodiversity loss, habitat fragmentation, competition with the food product, and wide-scale deforestation (Fridahl and Lehtveer, 2018). There is a specific factor of the uncertainty of sustainable biomass production to meet future energy needs. With increasing population, there is a demand for energy and growing demand for biomass production simultaneously. With

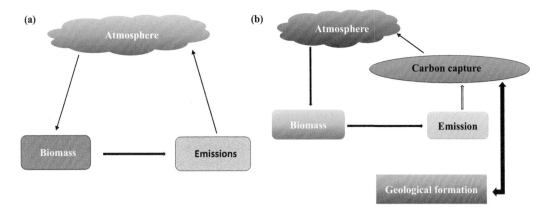

FIGURE 7.2 (a) Bioenergy carbon flow without BECCS. (b) Bioenergy carbon flow with BECCS.

the increase in demand, the demand for expansion of agricultural lands and loss of natural forests and eco-logical reserves will be there. Direct Land Use Change (LUC) and Indirect Land Use Change (iLUC) will exacerbate the greenhouse gas emissions from agricultural and forest systems. (Miyake et al., 2012; Kluts et al., 2017; IPCC, 2011). Although the future technical potential of BECCS is estimated to be considerable, it is not economically viable as, according to IPCC (2011:2007), each ton of carbon costs around 60–250 dol-lars. Large-scale employment of BECCS is a water-intensive process to obtain the maximum potential (Fazio and Monti, 2011). Due to the immense pressure on water resources, there will be impacts on the hydrological cycle, i.e., precipitation pattern, soil moisture loss, and water scarcity (FAO, 2014).

7.8 Biofuel for Mitigating the Future Climate Change

In the 21st century, climate change coupled with increasing energy demand, increase in oil price, and energy crisis have led to a search for an alternative energy source that will be efficient, renewable, economically viable, socially equitable, and environmentally sustainable(Owusu and Asumadu, 2016). Therefore, one of the energy sources with a positive response that has raised enormous response worldwide is 'Biofuel'. According to IEA (2010), energy demand will grow by 37% by 2040 globally. Renewable energy technologies play a dual role, i.e., to meet the world's energy demand and reduce greenhouse gas emissions (Schiermeier et al., 2008). The 'biofuels' in this chapter are the energy-enriched chemicals generated through the biological processes or derived from the biomass of living organisms (microalgae, plants, and bacteria). Biofuels are expected to meet the energy demand for at least one-fourth of the globe in the mid of 21th century (Ganesan et al., 2020).

7.8.1 Biofuel from Microalgae: A Potential Energy for the Future

The scientific community has done tremendous research on the first and second generations of biofuels. They were controversial due to their environmental implication, competition for food production, intensive land use, and expensive production processes. As a secondary alternative, microalgae species are found to be suitable for liquid biofuel production. Biofuel derived from microalgae is the third generation of biofuel (Bhagea et al., 2019). Microalgae are unicellular and multicellular photosynthetic microorganisms. Their simple structure facilitates a high growth rate and is proven to be photosynthetically efficient. Studies have shown that microalgae's biomass productivity could be 50 times more than that of switch grass, i.e., the fastest-growing terrestrial plant (Nakamura, 2006). Traditionally, terrestrial plants are not very efficient in capturing solar energy. There is an estimation that switch grass can convert solar energy to biomass energy at a yearly rate of $1\,W/m^2$, less than 0.5% of the solar energy received at a typical mid-latitude region (200–$300\,W/m^2$) (UNDP, 2015; Lewis and Nocera, 2006). On the other hand, studies have shown the microalgal photosynthetic efficiency range around 10%–20% or higher (Huntley and Redalje, 2007; Richmond, 2000).

7.8.2 Biohydrogen

Molecular hydrogen production by photosynthetic microorganism culture is one of the most promising renewable energy generation forms (Kruse et al., 2005). As the generation of molecular hydrogen through photosynthetic processes results in net-zero emission of GHG's and other environmental pollutants, there are projections that future efforts may lead to climate change mitigation by providing an environmentally friendly platform for industrial production of renewable energy. When biohydrogen is used to power fuel cells for electricity or in internal combustion engines, in both cases, the by-product is always water. Hydrogen does not produce harmful GHGs upon combustion but only water. Therefore, it is considered one of the energy sources to have the potential to replace conventional fossil-based fuels (Christopher and Dimitrios, 2012; Osman et al., 2020). However, there are challenges and difficulties for sustainable biohydrogen production. The literature mentions the possible drawbacks of hydrogen production (Allahverdiyeva et al., 2014); a few of them are as follows:

a. The photosynthetic rate of cell biomass and its further conversion to biohydrogen is low.
b. Under different solar light intensities, there is variation in the efficiency of light utilization by phototrophs.
c. The high level of ambient oxygen impairs the hydrogen production of enzymes and the pathways in cells
d. The current status of biohydrogen is not applicable because of the inadequate large-scale equipment for sustainable production.

7.8.3 Biodiesel

Biodiesel produced from renewable biological materials has received worldwide momentum as it can substitute the traditional petroleum diesel fuels. The transesterification of various animal fats and vegetable oils usually with methanol or ethanol generates biodiesel fuel (Naik et al., 2010). Some favourable aspects of biodiesel production are: It provides a cheaper, sustainable, and renewable alternative energy, and it is environmentally friendly as it recycles carbon dioxide (Cadenas and Cabezudo, 1998; Khan et al., 2009). The engine performances of non-toxic chemicals and combustion do not release sulphur and nitrogen-rich flue gas (Das and Roy, 2016). For the second generation of biodiesel production, an economically significant plant named *Jatropha curcas* has been identified as an appropriate candidate due to its unique characteristics such as high seed productivity and rapid growth. It can also grow in both tropical and subtropical zones of the world. Research studies have shown that microalgae can produce biodiesel 200 times more efficiently than traditional crops (Schenk et al., 2008).

7.8.4 Bioethanol

Ethanol is currently the most successful sustainable biofuel next to biodiesel (Liu et al., 2019). Bioethanol production mainly involves the fermentation of carbohydrates accumulated inside microalgae cells (Varfolomeev and Wasserman, 2011; Martín and Grossmann, 2013; Mint and Hanh, 2012). The productivity of third-generation bioethanol is significantly higher than other generations of biofuel (Mussatto et al., 2010). It has positive implications because it does not use arable land and reduces environmental impacts by sequestration of CO_2 from the atmosphere (Bastos, 2018).

7.8.5 Constraints in the Production of Biofuel from Microalgae

Generally, algal biofuels possess no or minimal negative impact on the environment. In fact, the biofuel production process could be associated with other environmental applications, including bioremediation (wastewater treatment), electricity or heat production, biofixation (CO_2 removal), biofertilizer, animal fodders, healthcare, and food products (Adeniyi et al., 2018). Algae are the most sustainable fuel feedstock that can help in decreasing greenhouse gases. However, it has certain drawbacks, and those are summarized in Figure 7.3.

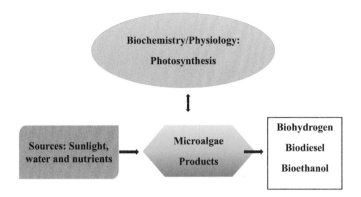

FIGURE 7.3 Contribution of microalgae in the energy sector.

7.9 Conclusion

The efficient use of biomass energy or (bioenergy) for mitigating future climate change depends on several underlying factors, some of which are difficult to quantify due to their degree of uncertainty. There is a need for an energy-efficient solution to meet the civilization's energy needs in a fast-growing world with the potential for future climate change. Among the conventional and non-conventional energy sources, the potential of biomass energy's utilization has gained momentum in recent years. It has often been looked upon as a promising renewable energy source to mitigate future climate change. The future of biomass energy depends on the efficient use of technologies to reach the climate goals. The bioenergy technology which is at the forefront is BECCS. There are always two sides to scientific temperament. On one side, some literature highlights the significant contribution of bioenergy in climate change efforts by deploying BECCS, its implication on phasing out fossil fuels, and environmental management.

On the other hand, literature is also directed at the discussion towards its risk in the large-scale deployment of BECCS, the controversy over food vs. fuel, and reduction of the land for carbon sequestration. Due to the future projections of an energy crisis and controversies surrounding food vs. fuel, scientists and researchers extracted biofuel from microalgae. Biofuel from microalgae is a third-generation biofuel and does not compete for food production. However, the story is different on the other side as it has certain negative drawbacks. It is not viable economically. Therefore, cost-effective technologies must be identified for large-scale production. There are barriers in global deployment levels, uncertainties in technological advancement, and obscurity in the policymaker's decision-making. However, bioenergy's role in climate mitigation overshadows the negative drawbacks. There is a need for robust research and development to keep the possible bioenergy option at the forefront. It promises alternative energy that will be a potential fuel to overcome the future energy crisis.

REFERENCES

Abbasi, T. and S. A. Abbasi. 2010. Biomass energy and the environmental impacts associated with its production and utilization. *Renewable and Sustainable Energy Reviews* 14(3):919–37.

Adeniyi, O. M., U. Azimov and A. Burluka. 2018. Algae biofuel: Current status and future applications. *Renewable and Sustainable Energy Reviews* 90, 316–335.

Allahverdiyeva, Y., E. M. Aro and S. N. Kosourov. 2014. Recent developments on cyanobacteria and green algae for biohydrogen photoproduction and its importance in CO_2 reduction. In *Bioenergy Research: Advances and Applications*, 367–87. Elsevier Inc. https://doi.org/10.1016/B978-0-444-59561-4.00021-8.

AR4-IPCC. 2007. Sustainable Development and Mitigation Coordinating: Mitigation of Climate Change.

Azar, C., K. Lindgren, E. Larson and K. Möllersten. 2006. Carbon capture and storage from fossil fuels and biomass - costs and potential role in stabilizing the atmosphere. *Climatic Change*. https://doi.org/10.1007/s10584-005-3484-7.

Bastos, R. G. 2018. Biofuels from microalgae: Bioethanol, green energy and technology, 229–246. https://doi.org/10.1007/978-3-319-69093-3_11.

Bhagea, R., V. Bhoyroo and D. Puchooa. 2019. Microalgae: The next best alternative to fossil fuels after biomass. *Microbiology Research* 10(1). https://doi.org/10.4081/mr.2019.7936.

Bilgen, S., S. Keles and K. Kaygusuz. 2007. The role of biomass in greenhouse gas mitigation. *Energy Sources, Part A: Recovery, Utilization and Environmental Effects* 29(13): 1243–52. https://doi.org/10.1080/00908310600623629.

Cadenas, A. and S. Cabezudo. 1998. Biofuels as sustainable technologies: Perspectives for less developed countries. *Technological Forecasting and Social Change* 58:83–103. Elsevier Inc. https://doi.org/10.1016/s0040-1625(97)00083-8.

Cassman, K. G. and A. J. Liska. 2007. Food and fuel for all: Realistic or foolish? *Biofuels, Bioproducts, and Biorefining* 1(1). https://doi.org/10.1002/bbb.3.

Celia, M., B. Gunter, J. Ennis and E. Lindeberg. 2005. Underground Geological Storage. IPCC chapter 5.

Christopher, K. and R. Dimitrios. 2012. A review on exergy comparison of hydrogen production methods from renewable energy sources. *Energy and Environmental Science.* https://doi.org/10.1039/c2ee01098d.

Climate Change: Atmospheric Carbon Dioxide | NOAA Climate.Gov. 2020 https://www.climate.gov/news-features/understanding-climate/climate-change-atmospheric-carbon-dioxide. (Accessed March 19, 2021)

Costello, A., M. Abbas, A. Allen, S. Ball, S. Bell, R. Bellamy and S. Friel, et al. 2009. Managing the health effects of climate change. Lancet and university college london institute for global health commission. *The Lancet.* https://doi.org/10.1016/S0140-6736(09)60935-1.

Das, D. and S. Roy. 2016. Liquid fuels production from algal biomass. In *Algal Biorefinery: An Integrated Approach*, 277–96. Springer International Publishing. https://doi.org/10.1007/978-3-319-22813-6_13.

FAO. 2014. FAO Success Stories on Climate-Smart Agriculture. Fao.

Fazio, S. and A. Monti. 2011. Life cycle assessment of different bioenergy production systems including perennial and annual crops. *Biomass and Bioenergy* 35(12). https://doi.org/10.1016/j.biombioe.2011.10.014.

Fridahl, M. and M. Lehtveer. 2018. Bioenergy with carbon capture and storage (BECCS): Global potential, investment preferences, and deployment barriers. *Energy Research and Social Science* 42. https://doi.org/10.1016/j.erss.2018.03.019.

Ganesan, R., S. Manigandan, M. S. Samuel, R. Shanmuganathan, K. Brindhadevi, N. Thuy Lan Chi, P. U. Duc and A. Pugazhendhi. 2020. A review on prospective production of biofuel from microalgae. *Biotechnology Reports.* Elsevier B.V. https://doi.org/10.1016/j.btre.2020.e00509.

Global Status of BECCS Projects 2010- Global CCS Institute. n.d.. https://www.globalccsinstitute.com/resources/publications-reports-research/global-status-of-beccs-projects-2010/. (Accessed March 14, 2021).

Haines, A., A.J. McMichael, K.R. Smith, I. Roberts, J. Woodcock, A. Markandya and B.G. Armstrong. 2009. Public health benefits of strategies to reduce greenhouse-gas emissions: overview and implications for policy makers. *The Lancet.* https://doi.org/10.1016/S0140-6736(09)61759-1.

Hansen, S. V., V. Daioglou, Z. J. N. Steinmann, J. C. Doelman, D. P. V. Vuuren and M. A. J. Huijbregts. 2020. The climate change mitigation potential of bioenergy with carbon capture and storage. *Nature Climate Change* 10(11). https://doi.org/10.1038/s41558-020-0885-y.

Hare, B. and M. Meinshausen. 2006. How much warming are we committed to and how much can be avoided? *Climatic Change.* https://doi.org/10.1007/s10584-005-9027-9.

Huntley, M. E. and D. G. Redalje. 2007. CO_2 mitigation and renewable oil from photosynthetic microbes: A new appraisal. *Mitigation and Adaptation Strategies for Global Change* 12(4). https://doi.org/10.1007/s11027-006-7304-1.

IEA. 2017. Bioenergy and Biofuels. http://www.Iea.Org/Topics/Renewables/Bioenergy/ n.d. (Accessed March 14, 2021).

Intergovernmental Panel on Climate Change. 2007. Sustainable Development and Mitigation Coordinating: Mitigation of Climate Change. https://www.ipcc.ch/site/assets/uploads/2018/03/ar4_wg3_full_report-1.pdf

Intergovernmental Panel on Climate Change. 2011. Renewable energy sources and climate change mitigation. https://www.ipcc.ch/report/renewable-energy-sources-and-climate-change-mitigation/

Intergovernmental Panel on Climate Change. 2012. Issues related to mitigation in the long term context. In Climate Change 2007. https://doi.org/10.1017/cbo9780511546013.007.

Kaygusuz, K. 2010. Climate change and biomass energy for sustainability. *Energy Sources, Part B: Economics, Planning and Policy* 5(2): 133–46. https://doi.org/10.1080/15567240701764537.

Khan, S. A., Rashmi, M. Z. Hussain, S. Prasad and U. C. Banerjee. 2009. Prospects of Biodiesel Production from Microalgae in India. *Renewable and Sustainable Energy Reviews.* Pergamon. https://doi.org/10.1016/j.rser.2009.04.005.

Klass, D. L. 2004. n.d. Biomass for Renewable Energy and Fuels. (Accessed March 14, 2021).

Kluts, I., B. Wicke, R. Leemans and A. Faaij. 2017. Sustainability constraints in determining european bioenergy potential: A review of existing studies and steps forward. *Renewable and Sustainable Energy Reviews.* https://doi.org/10.1016/j.rser.2016.11.036.

Kruse, O., J. Rupprecht, J. H. Mussgnug, G. C. Dismukes and B. Hankamer. 2005. Photosynthesis: A blueprint for solar energy capture and biohydrogen production technologies. *Photochemical and Photobiological Sciences* 4(12): 957–70. https://doi.org/10.1039/b506923h.

Kumar, A., P. Kumar, H. Singh and N. Kumar. 2020. Adaptation and mitigation potential of roadside trees with bio-extraction of heavy metals under vehicular emissions and their impact on physiological traits during seasonal regimes. *Urban Forestry & Urban Greening* https://doi.org/10.1016/j.ufug.2020.126900.

Kumar, A., S. Tewari, H. Singh, P. Kumar, N. Kumar, S. Bisht, S. Kushwaha, N. Tamta and R. Kaushal. 2021. Biomass accumulation and carbon stocks in different agroforestry system prevalent in Himalayan foothills, India. *Current Science* 120(6):1083–1088.

Lewis, N. S. and D. G. Nocera. 2006. Powering the planet: Chemical challenges in solar energy utilization. *Proceedings of the National Academy of Sciences of the United States of America.* https://doi.org/10.1073/pnas.0603395103.

Lindfeldt, E. G. and M. O. Westermark. 2008. System study of carbon dioxide (CO_2) capture in bio-based motor fuel production. *Energy* 33(2). https://doi.org/10.1016/j.energy.2007.09.005.

Liu, C. G., K. Li, Y. Wen, B. Y. Geng, Q. Liu and Y. H. Lin. 2019. Bioethanol: New opportunities for an ancient product. In Y. Li, X. Ge (Eds.), *Advances in Bioenergy*, Elsevier, 2019, 1–34. https://doi.org/10.1016/bs.aibe.2018.12.002.

Martín, M. and I. E. Grossmann. 2013. Optimal engineered algae composition for the integrated simultaneous production of bioethanol and biodiesel. *AIChE Journal* 59:2872–2883. https://doi.org/10.1002/aic.14071.

Miltenburger, C., F. Borgstrom and O. Eklund. 2018. Ecological benefits of health technologies – how to consider public value in funding decisions? *Value in Health* 21. https://doi.org/10.1016/j.jval.2018.07.188.

Minh, T. H. and V. Hanh. 2012. Bioethanol production from marine algae biomass: Prospect and troubles. *Journal of Vietnamese Environment* 3(1):25–29.

Miyake, S., M. Renouf, A. Peterson, C. McAlpine and C. Smith. 2012. Land-use and environmental pressures resulting from current and future bioenergy crop expansion: A review. *Journal of Rural Studies.* https://doi.org/10.1016/j.jrurstud.2012.09.002.

Mussatto, S. I., G. Dragone, P. M. R. Guimarães, J. P. A. Silva, L. M. Carneiro, I. C. Roberto, A. Vicente, L. Domingues and J. A. Teixeira. 2010. Technological trends, global market, and challenges of bio-ethanol production. *Biotechnology Advances* 28: 817–830. https://doi.org/10.1016/j.biotechadv.2010.07.001.

Naik, S. N., V. V. Goud, P. K. Rout and A. K. Dalai. 2010. Production of first and second generation biofuels: A comprehensive review. *Renewable and Sustainable Energy Reviews.* Pergamon. https://doi.org/10.1016/j.rser.2009.10.003.

Nakamura, D. N. 2006. Journally Speaking: The Mass Appeal of Biomass. Oil and Gas Journal.

NASA, Carbon Dioxide | Vital Signs – Climate Change: Vital Signs of the Planet. n.d. https://climate.nasa.gov/vital-signs/carbon-dioxide/. (Accessed March 14, 2021).

Obersteiner, M. 2001. Managing climate risk. *Science* 294 (5543). https://doi.org/10.1126/science.294.5543.786b.

Osman, A. I., T. J. Deka, D. C. Baruah and D. W. Rooney. 2020. Critical challenges in biohydrogen production processes from the organic feedstocks. *Biomass Conversion and Biorefinery.* https://doi.org/10.1007/s13399-020-00965-x.

Owusu, P. A. and S. Asumadu-Sarkodie. 2016. A review of renewable energy sources, sustainability issues and climate change mitigation. *Cogent Engineering.* Cogent OA. https://doi.org/10.1080/23311916.2016.1167990.

Read, P. and J. Lermit. 2005. Bio-energy with carbon storage (BECS): A sequential decision approach to the threat of abrupt climate change. In Energy. Vol. 30. https://doi.org/10.1016/j.energy.2004.07.003.

Renewable Energy Sources and Climate Change Mitigation — IPCC. 2021. https://www.ipcc.ch/report/renewable-energy-sources-and-climate-change-mitigation/. (Accessed March 14, 2021).

Richmond, A. 2000. Microalgal biotechnology at the turn of the millennium: A personal view. *Journal of Applied Phycology* 12. https://doi.org/10.1023/a:1008123131307.

Rosillo-Calle, F. 2016. A review of biomass energy - shortcomings and concerns. *Journal of Chemical Technology and Biotechnology.* https://doi.org/10.1002/jctb.4918.

Schenk, P. M., S. R. Thomas-Hall, E. Stephens, U. C. Marx, J. H. Mussgnug, C. Posten, O. Kruse and B. Hankamer. 2008. Second generation biofuels: High-efficiency microalgae for biodiesel production. *BioEnergy Research* 1(1): 20–43. https://doi.org/10.1007/s12155-008-9008-8.

Schiermeier, Q., J. Tollefson, T. Scully, A. Witze and O. Morton. 2008. Energy alternatives: Electricity without carbon. *Nature.* https://doi.org/10.1038/454816a.

Sillanpää, M. and C. Ncibi. 2017. Biomass: The sustainable core of bioeconomy. *A Sustainable Bioeconomy.* https://doi.org/10.1007/978-3-319-55637-6_3.

Sürmen, Y. 2003. The necessity of biomass energy for the turkish economy. *Energy Sources* 25 (2). https://doi.org/10.1080/00908310390142145.

The Paris Agreement I UNFCCC. 2016. https://unfccc.int/process-and-meetings/the-paris-agreement/the-paris-agreement. (Accessed March 14, 2021).

The State of Canada's Forests 2002–2003 I Canadian Forest Service Publications I Natural Resources Canada. n.d. https://cfs.nrcan.gc.ca/publications?id=22838. (Accessed March 14, 2021).

Varfolomeev, S. D. and L. A. Wasserman. 2011. Microalgae as a source of biofuel, food, fodder, and medicines. *Applied Biochemistry and Microbiology* 47:789–807. https://doi.org/10.1134/S0003683811090079.

Wilkinson, P., K. R. Smith, M. Davies, H. Adair, B. G. Armstrong, M. Barrett and N. Bruce. 2009. Public health benefits of strategies to reduce greenhouse-gas emissions: Household energy. *The Lancet.* https://doi.org/10.1016/S0140-6736(09)61713-X.

World Energy Assessment: Energy and the Challenge of Sustainability I UNDP. 2000 https://www.undp.org/publications/world-energy-assessment-energy-and-challenge-sustainability. (Accessed March 14, 2021).

World Energy Outlook 2010 – Analysis - IEA. 2010 https://www.iea.org/reports/world-energy-outlook-2010. (Accessed March 14, 2021).

8

Biofuel Production by Using Biomass and Its Application

Sarita Bisht, Amit Kumar, Narendra Kumar,
Hukum Singh, and Parmanand Kumar

CONTENTS

8.1 Introduction

The depleting reservoir of fossils and climate change has led to the rising demand for biofuel. Biofuel is defined as any hydrocarbon fuel generated in a short time period (days, weeks, or even months) through organic material (living or once-living). The UN (2008) has termed biofuel as "any liquid fuel made from plant material that can be used as a substitute to petroleum-derived fuel". Gaseous fuels produced from biomass-based sources have also been added to biofuels by the International Energy Agency (IEA Bioenergy, 2016). So, it can be understood that biofuels are liquid or gaseous components that are used as

DOI: 10.1201/9781003175926-8

fuels. India imports about 82% of its crude oil requirement from middle east, about 203 million metric tons of crude oil during 2016, which increased to 217 million metric tonnes during 2018 (Bacovsky et al., 2013).

According to IEA, the 2011 report, biofuels alone can cover up to 27% of global transport fuel requirements by 2050 to make full use of the fact that we need to make significant progress in technology and infrastructure. A lucrative international market has been created for industrial countries, among which Brazil and the United States are the leading ethanol producers. Countries such as the United States use corn (maize) grain to produce ethanol, which is mainly blended with gasoline to produce "gasohol," a fuel of 10% ethanol.

The results of several schemes and programs initiated by the Government to increase the blending of biofuels with positive results in this field can be traced from 2014. Many farmers have been encouraged to plant several species of plants to obtain biofuels. However, the plantations of *Jatropha* and *Pongamia* are not very productive. The yield of trees is low, although it is planted in poor soil and bears fruit. Biofuels are mainly produced from biomass rich in sugar and starch, such as sugar cane, sugar beet, sweet sorghum maize, and tapioca or oils such as soya, rapeseed, coconut, and sunflower (Ajanovic, 2011).

8.2 Biofuel Commercialization

Biofuel is commercialized mainly into two major forms (a) Primary biofuels and (b) Secondary biofuels. Primary biofuels are unprocessed wood fuel and are an example of organic matter utilized primarily in its natural state (as harvested) (Figure 8.1). Secondary biofuels are typically manufactured as solids (charcoal), liquids (alcohol, vegetable oil), or gases (e.g. Biogas, which is a combination of methane and other gases) (Figure 8.2). These two types of biofuels are largely traded energy carriers in formal and informal markets (Eide, 2008).

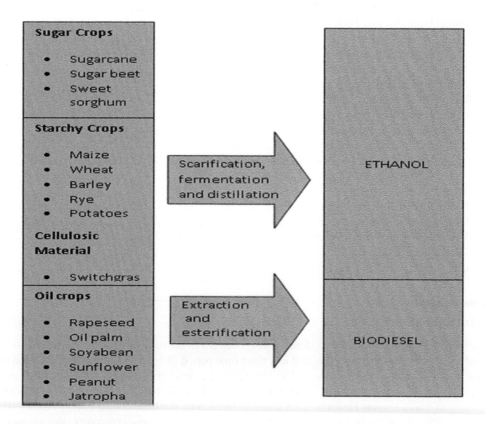

FIGURE 8.1   Biofuel crops, feed-stocks, and fuels.

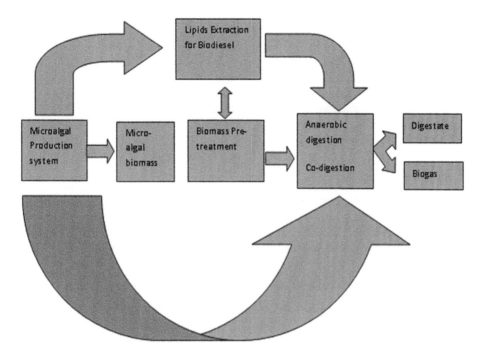

FIGURE 8.2 Anaerobic digestion of microalgae.

8.2.1 Classification of Biofuels

They are classified according to generations based on the type of feedstock used and the associated conversion method used. The following section discusses this classification in detail.

8.2.1.1 First Generation

It is procured mainly from grains or seeds primarily obtained from the plant's above-ground biomass's edible portion. Production involves a conventional well-established and relatively simple process (United Nations, 2008). Ethanol is made by fermenting sugars from starch-rich crops like sugar beet, sugarcane, corn and others (Martin, 2010).

8.2.1.2 Second Generation

Lignocellulosic biomass is the primary source of production of this type and can be obtained from feedstock grown in marginal arable and non-food crops and their residues (IEA Bioenergy, 2008). They can be produced from edible residues of food crop production e.g. corn stalks or rice husks and non-edible whole plant biomass e.g. grass or trees that are specifically grown for energy production (United Nations, 2008). They can be further classified into biochemicals and thermochemical based on conversion technology. Ethanol is the most common end product. However, further research is required for its competitive production without subsidies (IEA Bioenergy, 2016).

First- and second-generation biofuels have few limitations that prevent them from becoming a long-term alternative to petroleum. They can only be used in a small blend in the engine; otherwise, the engines must be modified. With the rising population and shortage of food crops and land the critical problem arises from food-based feedstocks, competition for available cropland, freshwater, and fertilizers (Kagan, 2010). Moreover, these fuels cannot be used in the Jet fuel market, a large transportation fuel segment (Sims et al., 2008).

8.2.1.3 Third Generation

The production of raw materials for third-generation biofuel neutralized concerns regarding competition of biofuel feedstock with food producers as they can be produced in non-arable land, out of which algae emerges as the most promising feedstock (IEA Bioenergy, 2010). Third-generation biofuel uses lesser resources in generating feedstock than second-generation biofuel, which used integrated technologies that involve the destruction of the whole biomass to produce feedstock and fuel (or fuel precursor, such as pure vegetable oil). Research has been widely carried out to solve problems related to production costs and metabolic production of fuels (Nanda et al., 2018).

8.2.1.4 Fourth Generation

The fourth-generation biofuels can also be produced efficiently without necessarily destroying the whole biomass in a non-arable land. This production technology focuses more on the use of renewable and widely available sources of energy such as solar energy. Photo biological solar fuel and electro-fuels are the most advanced biofuel currently under extensive research (Goldemberg, 2008).

8.2.2 Classification Based on Food and Agricultural Organization (FAO)

The basis of FAO classification of biofuel differs from conversion technology and is based on the nature of feedstock, its origin and the content of energy. They are classified mainly into three groups, namely wood fuels, agro-fuels and municipal by-products. This classification helps record biofuel trades, their important trade forms and production status and covers the origin of biomass. This helps to better understand the global biofuel commercial and underdevelopment market (Eide, 2008).

8.3 India's Most Significant Biofuel Categories

1. *Ethanol*: It is made from biomass that contains sugar or starch, as well as other cellulosic materials (Figure 8.4).
2. *Biodiesel*: Non-edible vegetable oils, acid oil, used cooking oil/animal fat and bio-oil are all used in its production and these are methyl or ethyl esters of fatty acid (Figure 8.3).
3. *Advanced biofuels*: It is made from non-food crops (such as grasses and algae), lignocellulosic, or hazardous effluent and residue streams that emit little CO_2 and reduce greenhouse gas emissions when competing with food crops for space.
4. *Drop-in fuels:* Biomass, agricultural residues and contaminants such as Municipal Solid Waste (MSW), Plastic wastes and Industrial wastes among others are used to make it.
5. *Bio-CNG*: It is made from bio-gas which has a similar energy potential and structure to natural gas derived from fossil fuels.

By 2030, the Government of India's National Policy on Biofuels-2018 proposes 20% ethanol in gasoline and 5% biodiesel in diesel.

8.4 National Policy on Biofuels and Its Salient Features

The Policy identifies biofuels as "Basic Biofuels" and works to enable the extension of appropriate financial and social incentives, categorizing them into the following generations:

First Generation (1G) *viz.,* bioethanol, biodiesel and "Advanced Biofuels."
Second Generation (2G) *viz.,* Drop-in fuels include ethanol and municipal solid waste (MSW).
Third Generation (3G) *viz.,* bio-CNG, biofuels etc.

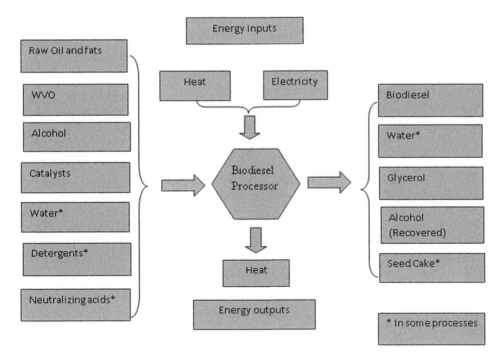

FIGURE 8.3 Process of biodiesel production.

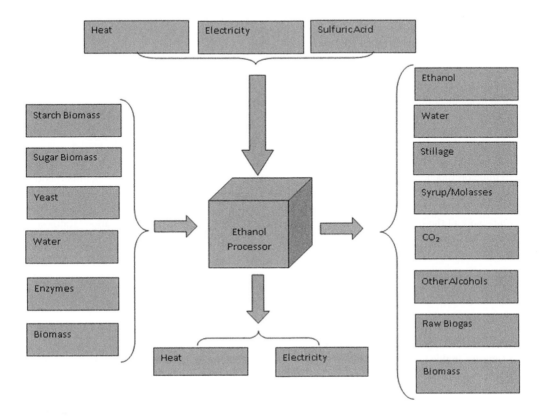

FIGURE 8.4 Process of ethanol production.

The Policy took the initiative to widen the scope of raw material used for ethanol production and extend financial assistance for the mass-established production of ethanol. With the approval of National Biofuel Coordination Committee, it has allowed surplus food crops such as (Lehrer, 2010).

- Starch-containing materials like Corn, Cassava
- Sweet sorghum, Sugar beet and sugarcane juice are examples of sugar-containing materials.
- For ethanol processing, damaged food grains such as wheat, broken rice and rotten potatoes are used.

The policy focuses on using non-edible oilseeds, used cooking oils and short-lived crops to create supply chain mechanisms for the production of biodiesel. It also offers a Rs. 5,000 crore viability gap financing scheme for 2G ethanol biorefineries for more than a 6-year period which includes a tax incentive and a higher purchase price over 1G biofuels. Rajasthan is the first state to implement the National Biofuels Policy. The State Government has released the Biofuel Rules for 2018. Highlights of the National Biofuel Policy of Rajasthan are:

- The State Government has focused on increasing the production of oilseeds
- Extended research in the fields of alternative fuels and energy resources
- The state has already installed a biodiesel plant with a production capacity of eight tons per day.
- The State Government has given further importance to the advertising of biofuels and to raise awareness for them the Women's Self-help Groups (SHGs) have been set up by the State Rural Livelihood Development Council to encourage the use of biofuels.
- The state government is also willing to help farmers dispose of their surplus stocks in an economical way that would reduce the country's dependence on oil imports.

The Government of Haryana and the Indian Oil Corporation (IOC) have also signed a Memorandum of Understanding (MoU) to establish an ethanol plant in the village of Bohali in the Panipat district with over Rs. 900 crores, which is valid for 1 year. This benefits farmer living within their 50 km radius area and helps meet the demand for ethanol and bio-fuels for consumers of petroleum products. This is intended to make proper use of crop residues and to reduce the amount of stubble farming in the area.

India's Ministry of Petroleum and Natural Gas (MoP&NG) has recently joined the IEA Bioenergy TCP to become its 25th member that was chosen with the main objective of easing the market launch of biofuels production in order to minimise emissions and crude oil imports. The Technology Collaboration Programme on Bioenergy (IEA Bioenergy TCP) of the International Energy Agency is an international forum for countries to enhance cooperation and knowledge sharing between countries with national bioenergy research, production and deployment programmes. The IEA Bioenergy TCP operates within the framework of the International Energy Agency (IEA), to which India has "Association" status since 30 March 2017.

8.5 Biofuel Programs and Their Impact in Rural India

8.5.1 World Biofuel Day

On 10 August 1893, Sir Rudolph Diesel (inventor of the diesel engine) succeeded in running a mechanical oil engine. In this regard, World Biofuels Day has been celebrated every year since 2015 on 10 August by the Ministry of Petroleum & Natural Gas of the Union. It can be said that this research experiment paved the way for vegetable oil to replace fossil fuels with different mechanical engines in the next century. On 10 August 2019, the Ministry of Petroleum & Natural Gas organized World Biofuel Day in Vigyan Bhavan, New Delhi, with the following objectives:

Objective: To raise knowledge about the importance of non-fossil fuels (also known as "environment-friendly fuels") as an alternative to traditional fossil fuels and to illustrate the government of India's initiatives in the biofuel field.

8.5.2 Other Initiative of Government of India to Promote the Use of Biofuels

Since 2014, the Government of India has taken several initiatives to increase the blending of biofuels.

1. The Government had taken steps to simplify the procurement procedures of OMCs and administer price mechanisms for ethanol. It has also amended the industries (Development & Regulation) Act, 1951enabling the lignocellulosic route for ethanol procurement.
2. The Government approved the National Biofuels Policy-2018 in June 2018 to achieve 20% ethanol blending and 5% biodiesel blending by the year 2030.
3. Among other things, the policy expands the feedstock scope for ethanol production and has provided incentives for producing advanced biofuels.
4. The price of C-heavy molasses-based ethanol has also been raised by the Government.

The Indian Renewable Energy Development Agency (IREDA) and Small Industries Development Bank of India (SIDBI) collaborated to create biodiesel plants, oil extraction/extraction units and infrastructure. The National Bank for Agriculture and Rural Development (NABARD) would provide refinancing for plantation loans to farmers. IREDA, SIDBI and other funding agencies act at different levels in the biofuel development process, commercial banks will be actively involved in funding multiple activities. The Biofuels Steering Committee must be established under the National Biofuels Co-ordinating Committee under this policy.

The Bio-Diesel Association of India (BDAI) is a national non-profit association representing the biofuels sector, specifically the biodiesel industry as the coordinating body for marketing, research and development in India. It also encouraged biodiesel-based biofuels and ensured sustainable agricultural growth, rural development, energy security and equal opportunities for the masses with overall environmental protection. Bioethanol has a concessionary excise duty of 16% under India's biofuel scheme, which exempts the biofuel industry from central taxes and excise duties. Plant and equipment for biodiesel processing are also eligible for discounts, and bioethanol concessions have also been provided on customs and excise duty. However, the policy continues to impose sales tax, license fee, permit fee and import taxes, which tend to hinder the industry's growth and development (Pahl, 2008).

The Ministry of Rural Development has taken a crucial step for the plantation of Jatropha/*Pongamia* seedlings by providing INR 490 million in financial assistance to nine identified states in 2005–2006 and INR 495 million to 15 states in 2006–2007 (IEA Bioenergy, 2016).

The state of Chhattisgarh's Department of Rural Development also used Mahatma Gandhi National Rural Employment Guarantee Scheme (MGNREGS) for planting TBO's (Tree Borne Oilseeds). The Joint Forest Management Committees (JFMCs) of Chhattisgarh have took several initiatives to plant biofuel crops in its forest land (Figure 8.1). Moreover, the Chhattisgarh Biodiesel Development Authority (CBDA) has also extended the plantation industry on forest land and wasteland by supplying Jatropha seedlings at highly subsidized rates. Participation of private investors and Gram/Intermediate Panchayats has also been encouraged to promote Jatropha cultivation and create village-level facilities for bio-oil extraction and sale route to biodiesel processing units. Minimum Support Price (MSP) for oilseeds has also been created in consultation with concerned government agencies, states and other stakeholders. MSP of Jatropha seeds was set at INR 6/kg by the governmental seed procurement agencies for farmers and higher INR 7–10/kg for private companies (Ravindranath et al., 2011).

The Government of Odisha provided local societies, such as Pani Panchayats and self-help organisations, 50% of the subsidies, while farmers above the poverty line received 33%, and 50% was given to farmers as groups under the funding of the coordination of Odisha Forest Development Corporation (OFDC) and Odisha Renewable Energy Development Agency (OREDA). The policy also allows the potential of interlinking the biofuel programme with other government programmes such as the Swarna

Jayanti Gram Swarozgar Yojana, MGNREGS, Integrated Tribal Developmental Agency, Compensatory Afforestation and Backward Regions Grant Fund.

Primary agricultural cooperative banks in the Tamil Nadu government are enabling subsidized loans to farmers for Jatropha cultivation on wasteland; 50% subsidy is offered for considering the agro-processing industries and plantations of jatropha; other biofuel crops are being established. The state emphasized contract farming of Jatropha by farmers with AGNI NET Biofuels Pvt. Ltd,

Mohan Bio Oils Ltd. and AHIMSA private companies. These companies offered Jatropha seed buy-back agreements with a market-linked price range of INR 5–10/kg. The Tamil Nadu Biofuel policy promotes the cultivation, production and usage of biofuels crops and biofuels, intending to achieve sustainable development.

8.5.2.1 Outcomes

1. These interventions of the Government of India have shown positive results.
2. Ethanol blending in petrol has reached an estimated 141 crore litres in the ethanol supply during the year 2017–2018from 38 crore liters in the year 2013–2014.
3. Bio-diesel blending in the country started from 10th August 2015, and in the year 2018–2019, Oil Marketing Companies have allocated 7.6 crore liters of biodiesel.
4. To address environmental issues arising out of the burning of agricultural biomass, the Oil PSUs have also agreed to set up 12 second-generation (2G) bio-refineries to augment ethanol supply.

Numerous ministries took an active role in the biofuel sector's advancement, growth and policymaking.

- The Ministry of New and Renewable Energy is in charge of overall strategy, encouraging the growth of biofuels as well as biofuel research and technology development.
- The Ministry of Petroleum and Natural Gas is in charge of selling biofuels as well as designing and enforcing pricing and procurement policies.
- The Ministry of Agriculture also played a key role in supporting biofuel feedstock crop research and development.
- Jatropha plantations on wastelands are to be promoted by the Ministry of Rural Development.
- The Ministry of Science and Technology promoted biofuel crop research, especially in the field of biotechnology.

The Prime Minister's National Biofuel Coordination Committee (NBCC) was also formed to provide high-level coordination between various departments and agencies involved and review policy advising on various aspects of biofuel production, promotion and use.

8.6 Major Sources for Biofuel Production

a. *Cellulose:* Cellulose content is found in plant materials ranging from switchgrass to trees, such as hybrid poplar and willow. Industrial by-products like corn stalks could be used for cellulose fuel as well. Cellulose, being a fibre, must be broken down into sugar before conversion to fuel. Cellulose is abundant in nature and can be easily regrown and harvested, making it one of the cleanest burning materials.

b. *Algal Oil:* Algae are grown in water and have the advantage over the competition for land with other crops, unlike other biofuel sources. The common form of algae includes seaweed and pond scum. Algae's oil content amounts to 50% of their body weight and can be used for ethanol production. Algae has high productivity and overcrowd each other very fast, which could have both positive and negative effect. On the one hand, it enables potential supply for the high demand of fuel on a long-term basis and on the other hand, overcrowding can prevent photosynthesis and cause massive die-offs. Despite its productivity, there has been no technology to maintain algal growth, which has caused a huge drawback in commercial production.

c. *Corn:* Corn is a suitable feedstock for biofuel production (Figure 8.1). The most significant source of biofuel production in the United States is corn, and it is more sustainable than petroleum. It suffers from wide disagreement on the split between use of food and for fuel production. Corn is considered a dietary staple and researchers are hesitant to use traditional food crops extensively for fuel. However, after ethanol production, the by-product of distiller corn can still be fed to animals.

d. *Soy:* Soy has been a popular biofuel and is commonly used as biodiesel and jet fuel. Oil is extracted by squeezing soy through transesterification, which is an easy and inexpensive process (Figure 8.1). Like any other agricultural crop, there is a debate on the use of soy. The seed oil that goes into a gas tank could have gone to someone's stomach and it might prove challenging to say that one destination is more valuable than the other. However, the content of biofuel material is not as high as that in corn, sugar cane, cellulose, or algae.

e. *Camelina and* Jatropha: *Camelina* and Jatropha are plant-based biofuel feedstocks that have recently come into play (Figure 8.1). They are flowering plants that can be grown in various locations in arid areas and wastelands. They have a huge potential with the added benefit of making the soil more fertile. Jatropha plantation however requires skill management in terms of what is essential for abundant fruiting.

f. *Rapeseed:* Rapeseed oil is also known as canola oil. It is widely available in Canada and the United States and has been observed to burn cleaner fuel than petroleum. It is cheap and easy to produce, but it requires a large plantation area to respond efficiently to current demand.

g. *Methane:* Stockholm and Sweden had already used methane to run the entire fleet of buses. US researchers have found that it burns cleaner fuel than most other fuels used. The decomposition of organic matter produces methane as food and compost by the microorganism (Figure 8.3). So, it is only likely that a large amount of methane is being produced in dumps and landfills worldwide. However, to implement methane as a vehicle fuel, most of the car designs would have to undergo some significant modifications and adaptations, including additional infrastructure for gas stations, which could cause a major drawback.

h. *Animal Fat:* Poultry, piggery and cattle farming are important aspects of human life. Animal fat could be a rare potential biofuel feedstock as it can be used as a fuel for cars and trucks. In addition to the fact that many industrial products use the same thing, collecting leftover fat from animal food products can be a problem in itself.

i. *Paper Waste:* Recycled paper, paper sludge from production and even sawdust from the early stages of paper processing have all been examined as possible biofuel sources. While producers would have to compete with paper companies—who use waste in their recycling programs—researchers believe that some waste could be diverted to fuel production. However, many of these materials may be more difficult than they are worth.

Currently, the main biomass sources used are sugar cane and corn to produce bioethanol and rapeseed for biodiesel production (Figure 8.1). Other common sources include sunflower seeds, soybean, canola, peanuts, Jatropha, coconut and palm oil for biodiesel and wheat, sugar beet, sweet sorghum and cassava for bioethanol. In the case of the biodiesel sector, oil palm produces the highest amount of fuel, and sugarcane yields the highest amount of fuel in the bioethanol sector (World Watch Institute, 2012). Research and experiment are still carried on various conversion technologies of the second generation of feedstock, such as cellulosic by-products of forestry and agricultural sectors such as grassy crops, woody plants, wood residues, stems and stalk well as municipal wastes.

Brazil is the world's leading producer of bioethanol, using sugar cane as its main feedstock. Brazil is the only country with a large-scale biofuel industry that has made bioethanol economical for consumers, satisfying 40% of its vehicle fuel needs (Figueroa, 2009). The United States is the second-largest

TABLE 8.1

Oil Yield Potential of Some Crops

S. No.	Crop Type	Oil Yield Potential (L/ha)	References
1	Micro algae	47.5–142.5	Ajanovic (2011)
2	Jatropha	6.0	Hannon et al. (2010)
3	Oil palm	2.0	Balat (2011)
4	Canola	1.25	Eide (2008)
5	Rapeseed	1.2	Hannon et al. (2010)
6	Sunflower	1.0	Ajanovic (2011)
7	Soyabean	0.5	Agriculture Organization (2008)
8	Corn	0.2	Vohra et al. (2014)

TABLE 8.2

Ethanol Production from Different Feedstock

S.No.	Source	Ethanol Yield (gal/acre)	Ethanol Yield (L/ha)	References
1	Corn stover	112–150	1,050–1,400	Amin (2009)
2	Wheat	277	2,590	Vohra et al. (2014)
3	Cassava	354	3,310	Fischer et al. (2009)
4	Sweet Sorghum	326–435	3,050–4,070	Gautam et al. (2019)
5	Corn	370–430	3,460–4,020	Vohra et al. (2014)
6	Sugarcane	536–714	5,010–6,680	Sims et al. (2010)
7	Sugar beet	662–802	6,190–7,500	Fischer et al. (2009)
8	Switch grass	1,150	10,760	Sikarwar et al. (2017)
9	Micro algae	5,000–15,000	46,760–1,402,900	Amin (2009)

producer of bioethanol made from corn grain and provides a little less than 2% of total automotive fuel needs. Brazil and the U.S. contributed about 71% of bioethanol production worldwide in 2004 (calculated from Clarke, 2006). Europe is the leading producer of biodiesel, with Germany accounting for 55% of biodiesel's total production in 2005, followed by France with a 15% share.

India is a developing country with predominant food scarcity. It must wisely choose a crop as the main source of biodiesel production based on the potential for oil yields of different edible and non-edible crops. Sunflower, soybean, cotton seed, canola/rapeseed, palm kernel, palm seed, mustard seed oil and rice are popular feedstocks (Figure 8.1). Poultry, beef and pork fats can all be used to make biodiesel (Figure 8.3). However, palm oil and animal fat are responsible for the formation of soap due to their high free fatty acid content, which has an adverse effect on downstream processing and reduces biodiesel yield (Raju et al., 2012) (Tables 8.1 and 8.2).

8.7 Biofuel Production

8.7.1 Production Techniques

Biofuel production techniques widely vary, depending on the type of raw material, efficiency level, production volume, surrounding situation, end-users' requirement, etc (Table 8.3).

8.7.2 Biofuel Conversion Processes

8.7.2.1 Deconstruction

Innovative biofuels (such as cellulose ethanol and renewable hydrocarbon fuels) are usually generated in a multi-step procedure. The rigid structure of the plant cell wall, which is made up of the biological

TABLE 8.3

Biofuel Production Techniques and Their Products

Raw Material	Technique	Product	Product Type
Vegetable oil and animal fat	Hydrotreatment	Biodiesel	Hydro-treated biodiesel
Algae	Fermentation, extraction and Esterification	Biodiesel	Algal biodiesel
Lignocellulosic material	Advanced hydrolysis and fermentation	Biomass-to-liquids (BTL): FischerTropsch (FT) diesel synthetic (bio) diesel	Synthetic biodiesel
Lignocellulosic material	Advance hydrolysis and fermentation	Cellulosic bioethanol	Bioethanol

molecules cellulose, hemicellulose, and lignin that are tightly bound together, must first be broken down. This can be accomplished in one of two ways: by breaking down at high temperatures or breaking down at low temperatures (Amin, 2009).

8.7.2.2 Deconstruction at High Temperature

Deconstruction of high-temperature uses extreme heat and pressure to break down solid biomass into liquid or gaseous intermediates using intense heat and pressure. In this direction, there are three main pathways:

- Pyrolysis
- Gasification
- Hydrothermal liquefaction.

During pyrolysis, in an oxygen-free climate, biomass is progressively heated to high temperatures (500°C–700°C). Heat causes biomass to decompose into pyrolysis vapour, coal and carbon. The vapours are cooled and condensed into a liquid "bio-crude" oil after the char is extracted. Gasification is a somewhat different method in which biomass is subjected to an elevated temperature (>700°C) with some oxygen present to create synthesis gas (or syngas), which is a mixture primarily composed of carbon monoxide and hydrogen. Hydrothermal liquefaction is the favoured thermal process when dealing with wet feedstocks including algae.

8.7.2.3 Low-Temperature Deconstruction

To break down feed materials into intermediates, low-temperature deconstruction usually employs biological catalysts such as enzymes or chemicals. First, biomass is pre-treated to break down the physical structure of plant and algae cell walls, allowing sugar polymers like cellulose and hemicellulose to be more easily accessed. Hydrolysis is the method of breaking down polymers into basic sugar building blocks, either enzymatically or chemically.

8.7.2.4 Upgrading

Intermediates including syngas, crude bio-oils, sugars and other chemical building blocks must be upgraded after deconstruction to create a finished product. Sugar or gaseous intermediates may be fermented by microorganisms such as cyanobacteria, bacteria and yeasts to produce fuel mixture stocks and chemicals. Sugars and several other intermediate streams like syngas and bio-oil for example, can be treated with a catalyst to extract undesirable or reactive compounds and boost storage and handling properties. Upgraded products may be commercially available fuels or bioproducts, or stabilised intermediates appropriate for finishing in a petroleum refinery or a chemical manufacturing plant.

Three Berkeley Lab researchers, including Corinne Scown of the Energy Technologies Area have recently received the Energy Achievement Award from the Joint Bioenergy Institute as part of a larger team (JBEI). Here's more from the announcement by JBEI: the award is designed to recognize the contributions of employees of the Department of Energy (DOE) to the Department's mission and the benefit of the United States. The JBEI team was recognized as a pioneer in developing biomass-derived ion liquids (bionic liquids) to enable efficient, flexible, scalable and economically viable single-pot conversion technologies to support the production of biofuels and co-products (Gomez, 2008).

The team members honoured by this award are Tanmoy Dutta, N.V.S.N. Murthy Konda, Corinne D. Scown, Blake A. Simmons, Seema Singh, Aaron M. Socha, Jian Sun, and Feng Xu. This team is a model of inter-institutional collaboration that JBEI has made possible, with about half of the group affiliated with Sandia (Dutta, Singh, Socha, Sun and Xu) and the other half with Lawrence Berkeley National Laboratories (Konda, Scown and Simmons). The members worked together to advance research and development on cellulosic biofuels by enhancing pre-treatment biomass's economic and environmental sustainability. The bionic liquid-enabled integrated one-pot process reduces annual operating cost by 40% and water use/wastewater generation by approximately 85%. It can reduce greenhouse gas emissions by as much as 50%–85% compared to conventional gasoline. With clear economic and environmental benefits, the one-pot bionic liquid process may represent a breakthrough technology in cellulosic biofuel development (Larsen et al., 2008).

Biofuel production is currently not commercially viable, partly because conventional diesel is heavily subsidized in India and its environmental impact is not reflected in the price. Biodiesel production would currently be feasible at a retail price of around 45–50 rupees per litre. But to give it a competitive edge over conventional fuel, the cost would have to be around 25 rupees a liter (Joshi et al., 2017). Because so many Indians rely on cheap fuel, there is no chance that policymakers will reduce fuel subsidies substantially anytime soon. Moreover, the plantation of Jatropha and *pongamia* are not very productive. Although these trees have been used for decades, suitable varieties' systematic cultivation still need to be researched. Although trees planted on poor soil bear fruit, yields are low. Both small farmers and large agricultural enterprises are, therefore, cautious about investing in biofuel plantations. However, it is expected that research and development will improve the size of the harvest and the seeds' oil content in the coming years.

Demand for biofuels will grow both nationally and internationally. Oil reserves are limited and energy demand will continue to rise so that new technology will gain a competitive advantage sooner or later. Industrialized countries have already established a lucrative international market. For example, the EU Biofuels Directive is targeting a 10% blend by 2020. India is a huge country and has many federal states. India's economic viability will not in itself mean that biodiesel production leads to inclusive and sustainable rural development, says Naveen Jain of the Indian Biodiesel Association. He says it is important to focus on how cultivation is organized and how value chains are structured.

Accordingly, various kinds of the value chain are emerging with directions. The three main types are listed in the following.

8.7.2.4.1 State-Driven Cultivation

Nearly half of India's land is under state control. Large areas, especially woodland, are degraded, neither used for farming nor forestry. According to the Government, 7.2 million hectares could be used for cultivating biofuel plants. The cost of seed, fertilizer, and labour, the whole investment risk, in other words, is borne in full by the Government. Some plantations are huge monocultures, others based on intercropping and yet others small plantations on public land. Typically, the committees in charge of the plantations are local and therefore state-owned cultivation is also serving decentralization of decision-making. Local people pick the crops and sell the harvest. In many instances, the jatropha harvest is gathered by village women, thus providing an additional source of income for landless labour. Large areas of degraded land are made productive again. But if the state interferes too much in the biofuel market, the sector's sustainable development can be impeded. The state government of Uttarakhand, for example, sought a deal with a single manufacturer, promising to sell all jatropha seeds grown on state-owned land for 3.50 rupees a kilo. But hardly anyone is prepared to grow biofuel trees for so little money. In this case, market dynamics have been hampered by a lack of demand-side competition and an ill-considered public–private partnership (Ramachandra et al., 2013).

8.7.2.4.2 Cultivation by Small Farmers

Small farmers plant oil seeds on private land. Some grow larger plantations on fertile soil. The problem is that less food can be produced as a result. Extensive cultivation on arable land is currently rare, as more money can be made from most other crops. But if energy prices rise or the productivity of biofuel plants rises, competition with food will become a matter of concern. In some countries—Tamil Nadu for example—the contract-farming of biofuel trees is already underway. Private investors guarantee relatively high harvest prices, thus stimulating production. However, from the perspective of development policy, additional income for farmers must be weighed against the risks of food scarcity. At present, the majority of small farmers are committed to cultivating Jatropha and *Pongamia* in hedgerow plantings, mixed crops or previously unused, degraded land. Therefore, they enjoy an additional income source that does not overwhelm food production (Rajagopal, 2008).

8.7.2.4.3 Commercial Cultivation

Private entrepreneurs are investing in large-scale plantations. They want to maximize profits as they bear the investment risk. In rural areas, commercial cultivation paves the way for investment both in agriculture and in processing enterprises. Trees are cultivated on private land, sometimes even making productive use of land held for speculative purposes. On the other hand, efforts were made in Chhattisgarh to lease government land to state-owned oil companies. Both models are creating new sources of employment for rural people. However, poor communities have been illegally using government land for subsistence and livestock farming purposes and are now at risk of losing their livelihoods. It remains to be seen how and whether the Government takes account of them and which form of land use benefits communities most in the long run (Axelsson et al., 2012).

Regardless of how biofuel plantations are organized, the harvest can be processed and utilized in two ways. If processing is centralized, the fuel will probably be sold on the international market or utilized in the national transport sector. An example of decentralized processing is Karnataka, where a decentralized network of oil mills already exists. The state government of Chhattisgarh has even embraced a strategy of setting up oil mills and esterification plants in all districts to promote the local use of biofuels. There are village-level models of producing electricity from biofuels. Such decentralized utilities improve power supply and the local people benefit. The risks are that poor, but illegal land users will be displaced and that biofuel plantations, once they become more portable, will crowd out food production.

So far, the three models described above have been applied to only a few hundred thousand hectares. A new law sets the target of mixing conventional diesel with 20% biodiesel. That would call for around 11 million hectares of land to be used for biofuel production—something India is a long way from achieving. Whether the oil-tree initiative will succeed is still not clear. Although there have so far been only a few isolated cases of Jatropha and *Pongamia* being grown instead of food crops, the Indian Government's initial enthusiasm has faded in light of the current debate on food prices.

The direction in which the biofuel industry develops in India will hinge crucially on the policies pursued at national and state levels. What is needed are intelligent regulations that promote competition and, at the same time, protect the poor and the landless. India has now reached the point where value-chain models have been operational long enough to appraise their economic, environmental and social sustainability. Several state governments have drafted sensible policies that need to be supported and renewed. Pragmatic approaches are more promising than the ongoing ideology-driven, polarising debate on biofuels.

The state of Andhra Pradesh has been a pioneer in the promotion of biofuels in India. In 2005, as part of a national government programme to support biofuels, a 5% ethanol blend was launched. The state government is attempting to meet energy demands by incorporating the production of biofuel feedstock from plants such as Jatropha into ongoing development projects. One of the benefits of plants like these is that they are ingrained in people's culture and memory, making them more readily accessible.

Andhra Pradesh has seen a surge in biofuel plantations. According to statistics, nearly 15,000 hectares (both *Jatropha* and *Pongamia*) of the 16,000 hectares in the 13 rain shadow districts (rain-deficient areas of the state) that the state government defined as suitable for growing these plants are currently in use as biodiesel plantations. By involving the local government's Gram Panchayat, small and medium farmers with wasteland were established in a participatory manner (a team that acts as the representative of all villagers, also referred to as the Village Sabha or Panchayat). The Rain Shadow Area Development Department has been appointed by the state government to continue providing policy support to the Department of Rural Development that promotes biofuel plantations in collaboration with local entrepreneurs and Gram Panchayats.

The National Rural Jobs Guarantee Scheme, a central government-sponsored programme with the primary goal of providing livelihood protection to rural households for at least 100 days of employment per year, is providing financial assistance. The state government has established public–private collaborations with entrepreneurs that provide *Pongamia* seedlings, technical expertise on plantation development, and seed procurement at a state-set minimum price (US $271 per tonne). Farmers are permitted to sell seeds on the open market if they can sell them for a profit.

Farmers can earn US$800–1,400 per hectare per year from mature *Pongamia* plantations of one hectare produce between 3 and 5 tonnes of seeds. Since no other crop can grow on these wastelands despite significant land levelling and reclamation, farmers saw growing the hardy *Pongamia* as a viable choice. Despite the fact that leaf gall (small bumps on leaves caused by mite infestation) has been recorded in *Pongamia* plantations, farmers do not spray pesticides because of the cost and the assumption that they do not harm the plantation. Farmers can also cultivate intercrops (http://www.intercrop.dk/General.htm) including black gramme, green gramme and castor, which fetch between $US 600 and 1,000 per hectare in a successful year. Raising wages for landless labourers has a direct connection to rural growth, in addition to providing extra income to farmers. According to the entrepreneurs' preliminary estimates, a *Pongamia* plantation can provide 66 human days of employment per hectare per year, excluding labour for intercrop production. Farmers are almost likely to take advantage of intercropping opportunities, so these plantations can produce even more jobs.

8.8 Biofuel Production Challenges

Challenges—However, due to constraints like very poor Jatropha seed yield, limited availability of wasteland, and high plantation and maintenance costs, biodiesel projects became unviable

- Trial results with High-Yielding Varieties (HYVs) of Jatropha for biodiesel production have not been satisfactory. Consequently, because of the limited availability of biodiesel and the volatile nature of its prices, the speed of blending had suffered a setback. It is reported that Jatropha occupies only around 0.5 million hectares of low-quality wastelands across the country, of which 65%–70% are new plantations of less than 3 years.

- Furthermore, one of the most challenging aspects of implementing the biodiesel program has been establishing large-scale Jatropha cultivation. Biodiesel crops have a longer gestation period (3–5 years for Jatropha), which results in a longer payback period and additional difficulties for farmers who do not have access to government assistance. According to an ICAR report, the Jatropha-based biodiesel production program faces several challenges, including slow planting development, inadequate processing and marketing infrastructure underdeveloped distribution channels.

- While favourable government policies and the vigorous participation of local communities and private entrepreneurs can sustain the program in the short term, it is equally important to have a sound long-term strategy at our disposal. The current course is not likely to be adequate in the long term given the present choice of feedstocks, technology status and available policy. To meet the country's potential bio-energy demands, a significant research focus on the production of second- and third-generation feedstocks is critical. According to the OECD-FAO

Agriculture Outlook 2018–2027, the demand for biofuels is shifting towards developing countries, which are increasingly putting in place policies that favour a domestic biofuels market. Like ethanol, demand for biodiesel is also expected to decline in the US and EU, driving down demand for vegetable oil as feedstock. Brazil, Argentina, Indonesia and other developing countries, riding on favourable policies will see a growth in biodiesel demand.

• Looking at the potential of biodiesel production in India, there is an urgent need to undertake research by public sector OMCs to achieve a higher yield of feedstock, developing short-duration crops and jatropha cultivation through planned varietal improvement programs, particularly in a few selected areas of the country to establish its viability. Finally, the principal changes in policy required are a multi-feed feedstock approach, an attractive incentive mechanism, both at the feedstock stage and biodiesel production stage and research and development for increasing the yield from the feedstock.

8.9 Achievements in Biofuel Production

Ethanol blending in petrol is an effective way of increasing domestic petrol availability, and for those all-around efforts need to be made to increase ethanol production. Fuels have caught global attention in the last decade. They are bio-based green liquid fuels that have been proven to be effective replacements for petroleum in the transportation field. Being environment friendly, bio-fuels like ethanol and bio-diesel can help us to conform to stricter emission norms. Globally, several policies have given a fillip to bio-fuel production, leading to an increase in ethanol and bio-diesel output. To promote biofuels in India, a National Policy on Bio-fuels was formulated by the Union Ministry of New and Renewable Energy in 2009.

In January 2013, the Union government launched the Ethanol Blended Petrol (EBP) program, which made it mandatory for oil companies to sell petrol blended with at least 5% of ethanol. The Government initiated significant investments in improving storage and blending infrastructure. The National Policy on Bio-fuels had set a target of 20% blending of bio-fuel by 2017. But the ethanol story has not yet succeeded in India. Let us examine why shortfall in Ethanol supplies oil firms initially launched the EBP in 2003 to blend ethanol with petrol.

From December 1, 2015 to November 30, 2016, 111 crore liters of ethanol were procured by the Oil Marketing Companies (OMCs), which would be sufficient for blending of only 3.5%. During 2016–2017, because of drought in Karnataka and Maharashtra, overall sugarcane and ethanol production reduced considerably and only 66.5 crore litres could be procured from suppliers. According to the Indian Sugar Mills Association (ISMA), sugar mills are set to more than double the supply of ethanol to fuel retailers for blending with gasoline in 2017–2018. Ethanol manufactures and OMCs finalized supply contracts for a record 1.4 billion liters during 2017–2018 (to realize 4% blending), compared with 665 million liters a year ago. OMCs, however, find it hard to locally procure the sugar by-product at the government-fixed rates as state governments have imposed heavy taxes on ethanol, widely used in the liquor industry. Sugar mills also prefer to sell ethanol to distilleries, where they get a better price and quicker deals.

According to Union government rules, 10% of ethanol extracted from sugarcane can be mixed with petrol. However, till now the Government had not achieved the target due to inadequate supply. Reasons for this include the non-uniform distribution of raw material throughout India and the lack of compulsory transportation and storage. Regulatory and policy approaches on excise duty, storage and transportation of ethanol, and pricing strategy of ethanol compared to crude oil are yet to be revised and implemented effectively.

In India, sugarcane molasses is the major resource for bio-ethanol production. The inconsistency of raw material supply is the major cause behind the sluggish response to blending targets. Due to the cyclical nature of sugarcane production, ethanol production varies as well, and there is no guarantee that optimal supply levels will be available to satisfy demand at any given time. The blending targets are partially successful in the years of surplus sugar production but unfulfilled when it declines. Drastic fluctuation in sugar cane farming and sugar milling pricing has resulted in mill owners being hugely indebted to farmers. Uttar Pradesh (UP)'s sugar mills have unpaid cane arrears due to falling prices and a market glut and are saddled with huge quantities of molasses. It is reported that currently, UP sugar mills have unsold molasses of more than 2.62 million tonnes (MT) that liquor manufacturers have not procured.

Permission is required for transferring molasses intending to produce ethanol, and such applications are pending with the state excise department, which needs to be processed urgently. As ethanol's domestic sourcing is continuously failing to achieve the target, there is a need to look at other alternatives. The National Biofuels Policy, 2018, seeks to widen the range of feedstock for ethanol production from the present sugar-molasses to other waste such as rural–urban garbage and cellulosic and lingo-cellulosic biomass, in line with the "waste-to-wealth" concept. The permissible feedstock includes sorghum, sugar-beet, cassava, decaying potatoes, damaged grain including maize, wheat, rice and, most importantly, crop residues such as wheat and rice stubble. This allows farmers to sell their surplus output to ethanol manufacturers when prices slump.

The Union government has allowed ethanol procurement from non-food feedstock besides molasses-like cellulosic and lingo-cellulosic materials, including the petrochemicals route. Given the consistent under-supply of domestic ethanol from traditional sources, oil PSUs are establishing 12 2G ethanol bio-refineries across 11 states of country, namely, Punjab, Haryana, Uttar Pradesh, Gujarat, Madhya Pradesh, Maharashtra, Karnataka, Odisha, Bihar, Assam and Andhra Pradesh. Media reports say that oil PSUs have entered into MoUs with state governments and technology providers for 2G ethanol bio-refineries. The foundation stone for one bio-refinery in Bathinda, Punjab has already been laid (Sims et al., 2010).

The approximate expenditure for raising each bio-refinery is around Rs. 800–1000crores, and it is expected that an amount of Rs. Oil PSUs will spend 10,000 crores in setting up these 12 bio-refineries. Second-generation ethanol is based on biomass such as wheat straw, rice straw, and crop stubble that can be converted into ethanol. It is more expensive than first-generation ethanol. However, by producing 2G ethanol, India can also address a major environmental issue like crop residue burning, which is causing horrific pollution in cities like Delhi.global outlook. According to the OECD-FAO Agriculture Outlook 2018–2027, the demand for biofuels is shifting towards developing countries, which are increasingly putting in place policies that favour a domestic biofuels market.

While declining demand for transport fuel could reduce demand for ethanol in the United States and the European Union main markets for ethanol, strong growth is expected in Brazil, China and Thailand, stimulated by favourable policies. In China, demand for ethanol could increase further with the implementation of its proposed new ethanol mandate. According to projections, 84% of the total additional demand for ethanol in the next 10 years will come from developing countries. In many countries, mandatory blending rules impose a minimum share of ethanol and bio-diesel to be used in transport fuel. India, for example, has a target of blending 10% ethanol with petrol by 2022 to cut dependence on imports. To achieve the target, 313 crore liters of ethanol is required. In a slew of decisions, the Government has started encouraging sugar mills to divert from sugar and boost ethanol production.

Conclusion: Ethanol blending in petrol is an effective way of increasing domestic petrol availability, and for those all-around efforts need to be made to increase ethanol production.

The technology for cellulosic and lingo-cellulosic biomass is still evolving, which needs to be upgraded and refined for commercial operation. There is also the danger of undue exploitation of the liberalized policy by existing sugar-based ethanol units. The industry may prefer to convert cane juice directly into ethanol without making sugar in the current scenario. Such a move would become an ecological disaster as sugarcane is a cost-intensive crop that consumes a lot of water, which the country can ill-afford to grow merely for biofuel production. This move needs to be discouraged and closely monitored. Unless the supply of ethanol can be increased from sources other than sugarcane, its use will not be widespread. The Government recently proposed blending methanol in petrol as another alternative, but again supply is a problem.

However, a politically motivated claim that biofuel can be harmful with plenty of disadvantages is attributed to some abuse of biomass production and misuse. For example, moving palm oil with fossil fuel-powered trucks and burning peat bogs to prepare biomass can significantly reduce greenhouse emissions. Other claims impose that biofuel production competes with food stock, although biofuel is not necessarily the addible product or vigorously agricultural-land planted. Such misleading assumptions and debates guided the European Parliament in January 2018 to propose an end to palm oil import and use by 2030. The EU's moves to cut imports on palm oil is a suspect of a cyclical move to protect the EU vegetable oil producers and fossil fuel-reliant industry. There are enough statistics and empirical data to dispute such claims and evaluate the importance of biofuels for the future at the expense of the insignificant disadvantages.

Because of how algae are grown and produced in most algal ponds, they are prone to attack by fungi, rotifers, viruses or other predators. Consequently, algal pond collapse is a critical issue that companies must solve to produce algal biofuels cost-effectively. The issue was identified as a critical component in the Department of Energy's National Algal Biofuels Technology Roadmap.

8.10 Affordable Aggregation of Feedstock

In our "Three reasons why waste is the king of renewable fuels, "we wrote:

1. *The feedstocks are available at fixed, affordable prices*: sometimes free, sometimes even transitionally available with a negative-cost tipping fee. And available in fixed, long-term supply contracts.
2. The odious sources are generally already aggregated for health or noxious reasons.
3. They are less subject to considerations such as the indirect land-use change plaguing energy crops and evoking few protects, if any, from environmental extremists.

8.11 Conclusion

The current global energy crisis has attracted significant worldwide attention towards biofuel energy. According to the report of the World Resources Institute, the quantity and demand of fuel consumed globally is expected to grow rapidly. It also warned that the use of fossil energy causes significant problems and harmful effects on the environment.

Renewable energy sources are critical to solving the world's energy crisis. Biofuels are an excellent example of renewable energy produced by biological organisms that reduce the country's dependence on fossil fuels. Photosynthesis is capable of increasing the amount of plant and algal biomass by using large-scale atmospheric carbon dioxide. Therefore, biofuels and biomass-derived fuels are based on photosynthesis and may be the key to achieving energy requirements that are eco-friendly and cost-effective. Much work has been done to improve the efficiency and effectiveness of biofuel production processes from biomass algae. The biofuel of the third generation must be free from the drawbacks of the first two generations. Many works aim to optimize the algae cultivation system: the open-air system and the photo bioreactor.

Since Indian crude oil production is almost stagnating to overcome the country's fuel demand, although various indigenous technologies for bioethanol and biodiesel production from different bio-based feedstocks have been developed, due to the unavailability of sufficient feedstocks, these bio-fuels' industrial production is still in its infancy. The central government, along with state governments, need to take more and more effective initiatives to ensure adequate availability of feedstock to the biofuel industry.

REFERENCES

Agriculture Organization. 2008. The State of Food and Agriculture 2008: Biofuels: Prospects, risks and opportunities. Food and Agriculture Organizer 38.

Ajanovic, A. 2011. Biofuels versus food production: Does biofuels production increase food prices? *Energy* 36(4):2070–2076.

Amin, S. 2009. Review on biofuel oil and gas production processes from microalgae. *Energy Conversion and Management* 50(7): 1834–1840.

Axelsson, L., M. Franzén, M. Ostwald, G. Berndes, G. Lakshmi and N. H. Ravindranath, 2012. Jatropha cultivation in southern India: Assessing farmers' experiences. *Biofuels, Bioproducts and Biorefining* 6(3): 246–256.

Bacovsky, D., N. Ludwiczek, M. Ognissanto and M. Wörgetter, 2013. Status of advanced biofuels demonstration facilities in 2012. A report to IEA Bioenergy task 39.

Balat, M., 2011. Potential alternatives to edible oils for biodiesel production–A review of current work. *Energy Conversion and Management* 52(2): 1479–1492.

Clarke, N. R., J. P. Casey, E. D. Brown, E. Oneyma and K. J. Donaghy, 2006. Preparation and viscosity of biodiesel from new and used vegetable oil. An inquiry-based environmental chemistry laboratory. *Journal of Chemical Education*, 83(2): 257.

Eide, A. 2008. The right to food and the impact of liquid biofuels (agrofuels). Rome, Italy: FAO 26–2.

Figueroa, M. J. 2009. The inclusion of Environmental concerns in the development of Biofuel Policies for Transport. In *Proceedings from the Annual Transport Conference at Aalborg University*, Vol. 16, No. 1.

Fischer, G., E. Hizsnyik, S. Prieler, M. Shah and H. VanVelthuizen. 2009. Biofuels and food security. Implications of an accelerated biofuels production. Summary of the OFID study prepared by IIASA.

Gautam, R., N. A. Ansari, P. Thakur, A. Sharma and Y. Singh, 2019. Status of biofuel in India with production and performance characteristics: A review. *International Journal of Ambient Energy*, 1–17. doi: 10.1080/01430750.2019.1630298.

Goldemberg, J. 2008. The challenge of biofuels. *Energy & Environmental Science* 1(5): 523–525.

Gomez, L. D., C. G. Steele-King and S. J. McQueen-Mason. 2008. Sustainable liquid biofuels from biomass: the writings on the walls. *New Phytologist* 178(3): 473–485.

Hannon, M., J. Gimpel, M. Tran, B. Rasala, and S. Mayfield, 2010. Biofuels from algae: challenges and potential. *Biofuels*, 1(5): 763–784.

IEA Bioenergy. 2010. IEA Bioenergy Task 32.

IEA Bioenergy. 2016. Cascading of woody biomass: Definitions, policies and effects on international trade, Task 40.

IEA Bioenergy.2008. From 1st-to 2nd-Generation Biofuel technologies Biofuel technologies. An overview of current industry and RD&D activities. IEA-OECD.

Joshi, G., J. K. Pandey, S. Rana, and D. S.Rawat. 2017. Challenges and opportunities for the application of biofuel. *Renewable and Sustainable Energy Reviews* 79: 850–866.

Kagan, J. 2010. Third and fourth generation biofuels: Technologies, markets and economics through 2015. GreenTech Media Research.

Larsen, J., M. Ostergaard Petersen, L. Thirup, H. Wen Li and F. Krogh Iversen, 2008. The IBUS process–lignocellulosic bioethanol close to a commercial reality. *Chemical Engineering & Technology: Industrial Chemistry-Plant Equipment-Process Engineering*-Biotechnology, 31(5): 765–772.

Lehrer, N. 2010. (Bio) fueling farm policy: The biofuels boom and the 2008 farm bill. *Agriculture and Human Values* 27(4): 427–444.

Martin, M.A. 2010. First generation biofuels compete. *New Biotechnology* 27(5): 596–608.

Nanda, S., R. Rana, P. K. Sarangi, A. K. Dalai and J. A. Kozinski. 2018. A broad introduction to first-, second-, and third-generation biofuels. In Sarangi, P., Nanda, S., and Mohanty, P. (eds.) *Recent Advancements in Biofuels and Bioenergy Utilization*. Springer, Singapore, 1–25.

Pahl, G. 2008. *Biodiesel: Growing a New Energy Economy*. Chelsea Green Publishing, Hartford.

Rajagopal, D.2008. Implications of India's biofuel policies for food, water and the poor. *Water Policy* 10(S1), 95–106.

Raju, S. S., S. Parappurathu, R. Chand, P. K. Joshi, P. Kumar and S. Msangi. 2012. *Biofuels in India: Potential, Policy and Emerging Paradigms*. New Delhi: National Centre for Agricultural Economics and Policy Research.

Ramachandra, T. V., M. D. Madhab, S. Shilpi and N. V. Joshi. 2013. Algal biofuel from urban wastewater in India: Scope and challenges. *Renewable and Sustainable Energy Reviews* 21: 767–777.

Rauch, R. and J. Hrbek, 2015. Country Report Austria. IEA Bioenergy Task 33.

Ravindranath, N. H., C. S. Lakshmi, R. Manuvie, and P. Balachandra. 2011. Biofuel production and implications for land use, food production and environment in India. *Energy Policy* 39(10): 5737–5745.

Sikarwar, V. S., M. Zhao, P. S. Fennell, N. Shah and E. J. Anthony. 2017. Progress in biofuel production from gasification. *Progress in Energy and Combustion Science* 61; 189–248.

Sims, R., M. Taylor, J. Saddler and W. Mabee. 2008. From 1st-to 2nd-generation biofuel technologies. Paris: International Energy Agency (IEA) and Organisation for Economic Co-Operation and Development.

Sims, R. E., W. Mabee, J. N. Saddler and M. Taylor. 2010. An overview of second-generation biofuel technologies. *Bioresourcetechnology* 101(6): 1570–1580.

Task IEA Bioenergy. 2016. Cascading of woody biomass: Definitions, policies and effects on international trade.

UN. 2008. Sustainable Bioenergy: A Framework for Decision Makers. United Nations.

Vohra, M., J. Manwar, R. Manmode, S. Padgilwar and S. Patil, 2014. Bioethanol production: Feedstock and current technologies. *Journal of Environmental Chemical Engineering* 2(1): 573–584.

World watch Institute, 2012. Biofuels for transport: global potential and implications for sustainable energy and agriculture. *Earthscan.*

9

Biomass Gasification Technologies for Sustainability of Future Energy Demand

Ratnakiran D. Wankhade

CONTENTS

9.1 Introduction

Biomass is a clean and eco-friendly renewable source of energy. The biomass is used to contribute to the world power supply of 10%–14% (Mc. Kendry, 2002), which shows that biomass for energy production has great scope globally. The main advantage of biomass over fossil fuels like natural gas, petroleum, and coal is that it provides a continuous feedstock supply. Also, fossil fuels have the drawback of a lack of continuous supply, high prices, and pollution of the environment by the greenhouse effect (Paula et al., 2013). It has been observed that biomass is a CO_2-neutral resource over its life cycle (Li et al., 2004a, b) and possesses zero CO_2 net emission energy (Mohammed et al., 2011)

On the other hand, compared with fossil fuel, it transfers CO_2 to the atmosphere by the burning processes that is trapped underground in the form of coal gas and crude oil. In the case of the combustion of biomass, the carbon, which is already present in a neutral state like wood, is added to the atmosphere. Hence, no new carbon is produced, which results in no new CO_2 added to the atmosphere. Since the amount of sulphur in biomass-derived fuel is slightly lower (negligible), it contains less ash and waste than fossil fuels. Hence, there is no contribution of biomass combustion in producing sulphur dioxide,

DOI: 10.1201/9781003175926-9

which is the main reason for acid rain. The ash produced from biomass combustion can also be used as soil additives in the farmer's field.

The world also faces a waste management challenge from biomass such as municipal waste, agriculture residue, and forest biomass. Converting this biomass into syngas would address the waste management problem and minimize the amount of waste landfills. The biomass conversion for energy production also has economic advantages because it does not affect the world price fluctuation and uncertainties of supply, which reduces the dependency and economic pressure of fossil fuel imports (Demirbaş, 2001). The main process of obtaining energy from biomass is direct combustion. There are four basic biomass conversion technologies: thermochemical, biochemical, agrochemical, and direct combustion. The application of energy produced from the biomass is in different forms like space heating, cooking, industrial processes etc. Anaerobic digestion and alcoholic fermentation are the most common biochemical conversions, while agrochemical processes primarily use mechanical extraction methods, such as rapeseed oil extraction. Thermochemical processes are divided into four types: gasification, pyrolysis, liquefaction, and supercritical fluid extraction. Mostly, thermochemical decomposition is used for all biomass types with low moisture and of a woody and herbaceous nature (Demirbaş,2001). The word biomass defines itself: "bio" means life, and biomass means biological material obtained from living organisms. Biomass is generally classified into five different types: Virgin wood (biomass from forest waste), Energy crops (crop specially grown for energy application), Agricultural Residue, Food Waste, and Industrial waste. Among them, the major contribution of biomass is from wood and wood waste, i.e., 64% (Jansen, 2012).

9.2 Gasification of Biomass

The gasification process is the process in which organic compounds are converted into gas and solid products. The gas produced is called syngas, and the solid part is obtained called char. The obtained syngas is a mixture of CO, H_2, CH_4, CO_2, light hydrocarbon, and heavy hydrocarbon. Syngas contains certain undesirable gases such as hydrochloric acid (HCL), sulfuric acid (H_2S), and inert gases such as nitrogen. The lowest heating value of syngas ranges from 4–13 MJ/Nm³ depending on the operating situation, feedstock, and gasification technology (Qian et al.,2013; Wu et al.,2014; Liu and Ji,2013),while the char ranges from 25 to 30 MJ/kg (4). Generally, the gasification processes include the four major stages as follows (Molino et al., 2016)

1. 9.2.1 Oxidation (Exothermic Stage)
2. 9.2.2 Drying (Endothermic Stage)
3. 9.2.3 Pyrolysis (Endothermic Stage)
4. 9.2.4 Reduction (Endothermic Stage)

One more step of tar decomposition can also be considered, which includes the formation of light hydrocarbon.

9.2.1 Oxidation

The oxidation step of the gasification process is essential for achieving the thermal energy needed for endothermic processes to the required value of the operating temperature. To oxidise the only part of the fuel, the oxidation reaction is carried out in the absence of oxygen in relation to the stoichiometric ratio. The reactions (Molino et al., 2016) that take places during oxidation are as follow:

$$C + O_2 \rightarrow CO_2 \, OH = -394 \, kJ/mol \, Char \, combustion \qquad (9.1)$$

$$C + 1/2O_2 \rightarrow COOH = -111 \, kJ/mol \, Partial \, oxidation \qquad (9.2)$$

$$H_2 + 1/2O_2 \rightarrow H_2O\,OH = -242\,kJ/mol\,Hydrogen\,combustion \tag{9.3}$$

9.2.2 Drying

The moisture in the feedstock is evaporated in this process, which requires proportional heat for evaporation due to the moisture content of the feedstock. When the temperature of the process exceeds 150°C, the drying process is considered complete, and the heat needed is usually met from another stage of the process (Hamelinck et al., 2004).

9.2.3 Pyrolysis

Pyrolysis is the thermochemical decomposition process in the absence of oxygen for a short time. In this process, the breaking of the chemical bond takes place, resulting in the formation of molecules with lower molecular weight. With the help of pyrolysis process, different fractions can obtained, i.e., solid, liquid, and gaseous fractions. For fluidized bed gasifiers, the solid fraction by weight is 5–10 wt, whereas for fixed bed gasifiers, it is 20–25 wt%, with a high heating value and carbon content (Li et al., 2004; Lv et al., 2004; Gómez-Barea et al., 2005; Roos, 2010). This fraction consists of char, which includes the ash and high carbon content. The liquid fractions obtained are called tars, which have different percentages by weight according to the types of gasifier beds. It is lower than 1 wt% for downdraft gasifier, 1–5wt% for bubbling bed, and 10–20wt% for updraft gasifier. The tars mainly consist of complex organic substances that are higher in amount at relatively lower temperatures (Carpenter et al., 2007). The output gaseous fraction is 70–80 wt% of the input fed material, that is a mixture of gases which are incompressible at room temperature (Schmid et al., 2012). The pyrolysis gas is primarily composed of carbon monoxide, hydrogen, carbon dioxide, and light hydrocarbons including methane as well as other carbon, C_2, C_3 hydrocarbons.

In contrast, inert gas and acids are minor constituents. The temperature at which the pyrolysis process takes place is extremely high, varying between 250°C and 700°C. The heat needed for this process comes from the oxidation stage of the process, which is endothermic. The chemical reaction (Widyawati et al., 2011) that occurs during the pyrolysis process is as follows:

$$Biomass \rightarrow\leftarrow H_2 + CO + CO_2 + CH_4 + H_2O(g) + Tar + Char\,(Endothermic)... \tag{9.4}$$

Since cellulose (50 wt%) is the key constituent of biomass, the chemical formula for cellulose ($C_6H_{10}O_6$) may be used to identify the feedstock (Shen et al., 2011).

9.2.4 Reduction

All of the products of the oxidation and pyrolysis stages are involved in the reduction process. The char and the mixture of gases react in this stage, resulting in the formation of syngas. The main chemical reactions (Molino et al., 2016) that take place during the reduction stage are as follows:

$$C + CO_2 \leftrightarrow 2CO\,OH = 172\,kJ/mole\,Boudouard\,reaction \tag{9.5}$$

$$C + H_2O \leftrightarrow CO + H_2OH = 131\,kJ/mole\,Reforming\,of\,the\,char \tag{9.6}$$

$$CO + H_2O \leftrightarrow CO_2 + H_2$$

$$OH = -41\,kJ/mole\,Water\,gas\,shift\,reaction \tag{9.7}$$

$$C + 2H_2 \leftrightarrow CH_4OH = -75\,kJ/mole\,Methanation \tag{9.8}$$

9.3 Gasification Technologies

The different types of reactors are categorized into the following based on the contact mode between the gasifying agent and the feed material, the rate of heat transfer and the mode of transfer, and residence time in the reaction zone.

1. 7.3.1 Fixed bed Reactors
2. 7.3.2 Rotary kiln reactor
3. 7.3.3 Entrained flow reactor
4. 7.3.4 Fluidized bed reactor
5. 7.3.5 Plasma reactor

9.3.1 Fixed Bed Reactors

The fixed bed reactor is further divided into type Updraft reactor and Downdraft reactor. The main difference between updraft and downdraft gasifier is the direction of solid and gas movement into the reactor. In an updraft gasifier, the solid material moves downward. In contrast, the produced syngas moves upward in relation to the gasification agent, while both the solid material and the released gas flow downward in a downdraft gasifier. Table 9.1 lists the benefits and drawbacks of each of these reactor types (Van der Drift et al., 2004; Higman and van der Burg, 2008; Belgiorno et al., 2003; Canabarro et al., 2013; Begum et al., 2013)

TABLE 9.1

Benefits and Drawbacks of Fixed Bed Reactor

Reactor Type	Benefits	Drawbacks
Fixed bed "Updraft"	• High thermal efficiency • Contact between the solid material and the oxidising agent should be good. • Can handle materials of different sizes • Can work with high-humidity materials • Dust and ashes entrainment are also reduced. • Simple construction	• Syngas with a high tar content • Tar's energy content is greater than 20%. • CO and H_2 output is low. • necessitates a follow-up procedure for the tar cracking • Loading and processing flexibility is restricted (the treated material should have properties homogeneous) • "Updraft" in the fixed bed • Can work with high-humidity materials • Starting difficulties are reduced, and temperature control is improved. • To prevent the creation of preferential paths in the fixed bed, mobile grates must be mounted. • According to catalysts, they might not even be accessible because syngas energy is lower than that needed for activation, necessitating the use of external energy. • Deactivation of catalysts by poisoning could be possible.
Fixed bed "Downdraft"	• Robust technology • There are no problems of scale-up • High carbon conversion • Low production of tar • Limited entrainment of ash and dust • High solid residence time • Simple construction • Reliable technology	• Low specific capacity • Inputs must be standardised in scale (pellets no larger than 100 mm) • On the grid, sintered slag is being trained. • Materials with a low moisture content are required. • Low moisture content materials are needed. • Loading and processing flexibility is restricted (the treated material must have the same characteristics) • Low heat transfer coefficient • Starting and maintaining the temperature is difficult. • Scale-up possibilities are limited.

9.3.2 Rotary Kiln Reactor

The rotary kilns reactor made up of a cylindrical chamber tilted at 1%–3%, which rotates slowly around its axis. In these reactors, the continuous stirring action of drum gas and solid contact takes place, due to which a new solid surface is exposed to the gasification agent. In this system, the residence time, heat, and matter exchanges are not more effective than other gasification technologies. In this system, the efficiency of solid–gas contact can be improved by installing the barrier within the drum that enhances the contact area with the gaseous stream and increases solid material handling. The substance to be gasified goes in the top of the reactor, while the oxidising agent goes in the bottom. Table 9.2 lists the advantages and disadvantages of this reactor (Iovane et al., 2013; Molino et al., 2013; Donatelli et al., 2010).

9.3.3 Entrained Flow Reactor

The operating temperature in this type of reactor is very high, about 1,300°C–1,500°C, with a 25–30 bar pressure. The raw material in this reactor can be used as dry feed and water slurries. The feed fuel size is 0.1–1 mm, and the injection of the gasifying agent happens in a co-current way. Pressurized power solid fuel is feed into the gasifier by pneumatic feeding, but slurries are atomized and subsequently served as pulverized solid fuel. These types of reactors are further divided into slagging and non-slagging. The ash escapes the reactor in the form of liquid slag in the first reactor, while slag is not formed in the second reactor. The maximum allowable ash content is 1%. It is observed that, usually, a pre-treatment based on torrefaction is required when fine particle biomass is used as feedstock for reducing the moisture content and bulk density. The advantages and disadvantages of these types of reactor are shown in Table 9.3 (Van der Drift et al., 2004; Higman and van der Burg, 2008; Couhert et al., 2009).

TABLE 9.2

Advantages and Disadvantages of Rotary Kiln Reactor

Reactor Type	Advantages	Disadvantages
Rotary kiln	• Low sensitivity to modifications in the fed's composition, humidity, or scale • Maximum flexibility in loading • High conversion • Suitable for waste that has the potential to melt • There are scaling issues to consider • Construction simplicity and high operational efficiency • Investment costs are lower	• Starting and temperature management are extremely difficult • The existence of movement parts, as well as their leakage and wear issues • Refractory consumption is very high • Heat exchange capability is insufficient • There is a lot of dust and tar in this place • Heat with a low efficiency • Process with a small amount of flexibility • High maintenance costs

TABLE 9.3

Advantages and Disadvantages of Entrained Flow Reactor

Reactor Type	Advantages	Disadvantages
Entrained flow reactor	• Fuel flexibility • Temperature uniformity • Carbon conversion rate is high • There aren't any scaling issues • Excellent ability to control the parameters of the process • Reactor residence time is limited • Tar content is extremely low • High-temperature slagging (vitrified slag)	• Oxidant standards are high • Product gas has a high level of sensible heat • Heat recovery is required to boost efficiency • Low efficiency of cold gas • Size and preparation supply must be reduced • System components, including the gasifier vessel refractory, have a short life span • Plants are expensive • High maintenance costs

9.3.4 Fluidized Bed Reactor

The fluidized bed reactor is divided into two types like bubbling fluidized bed reactor and re-circulating bed reactor. In the bubbling fluidized type reactor, a bed of sand as inert granular material is held in a state of fluidization. Through the bottom distribution grid, the gasification agent is fuelled bottom-up at a velocity of 1–3m/s.

The presence of gas bubbles continuously stirs the inert solid bed that acts like a liquid. The uniformity of conditions is ensured by the mobility of gas bubbles in both heat transfer and the exchange of matter. The grid region extent may significantly affect the heat and mass transfer between bed solids and gasification agent. On the other hand, in the re-circulating bed reactor gasification, there are two stages. The first stage consists of the bubbling fluidized bed, the heat required for the occurrence of gasification processes generated by combustion. In the second stage, the dragging of the solid's is carried out by the high-speed gas of 5–10m/s, where the gasification and pyrolysis processes occur. The solid re-circulated to the bubbling fluidized bed is separated by the cyclone fitted at the outlet of the reactor. Due to the high mass and heat transfer rate, high mixing capacity, catalyst using as part of gasifier bed, and constant temperature, the fluidized bed is the most promising gasification technology than the other technologies for tar conversion. The advantages and disadvantages of the fluidized bed are shown in Table 9.4 (Higman and van der Burg, 2008; Chhiti et al., 2013; Liu and Ji, 2013).

9.3.5 Plasma Reactor

An ionized gas stream with a temperature of up to 10,000°C is obtained in the plasma reactor with the arc's electric discharge application. Electrodes are installed in the flashlight, which is typically made of

TABLE 9.4

Advantages and Disadvantages of Fluidized Bed Reactor

Reactor Types	Advantages	Disadvantages
Bubbling fluidized bed	• High carbon conversion • Mixing at a high rate and gas-solid contact • Thermal loads that are extremely high • Excellent temperature control (temperature distribution along the reactor) • Can work with a variety of materials with varying properties • Excellent load and process flexibility • Appropriate for highly reactive fuels like biomass and pre-treated municipal waste • In the syngas, there is a low level of tar • Start-up, shutdown, and control are all made simple • Catalysts can be used, even on a large scale • There are no moving parts. • Excellent scalability	• Carbon loss in the ashes • Dragging of dust and ashes • Requirement for pre-treatment with heterogeneous materials • To avoid the phenomenon of bed defluidization, the process temperature must be kept relatively low (temperature lower than the softening point of the solid residues) • Dimensional restraints • High initial and ongoing investment and maintenance costs
Circulating fluidized bed	• Lower tar production • High conversions • Flexible load • Reduced residence times • Excellent scalability	• It is possible to cast the ashes • Carbon loss in the ashes • Size reduction and preparation supply are required (the solid material must be finely pulverised with dimensions less than 100mm) • Contact between solids and gases is restricted • Special materials are required • Technology is complex and difficult to manage. • Concerns about security • Expensive start-up and investment costs

copper, but carbon electrodes can also be used. In the presence of an oxidising agent, it is used to treat organic matrixes; the plasma process causes atomic degradation of the matter. The temperature rise required for the gasification reaction is determined by the energy flow supplied by the plasma, but it is similar to pyrolysis in the absence of an oxidizing agent. The plasma technology is used for the thermal treatment in two ways: In the first thermal destruction, process plasma is directly applied to the solid to be treaded of a specific size, allowing temperature control independent of the other parameters. In the second one, the goal is to maximize the synthesis of gas production with light components in a high amount and remove the tar present at the initial stage. The advantages and disadvantages of plasma arc welding are shown in Table 9.5 (Zhang et al., 2012).

9.4 Suitability of Biomass

The gasifiable biomass is made up of hemicellulose, lignin, cellulose, and proteins, some of which are listed in Table 9.6 (Cheng et al., 2014). The lignin behaves like the fibre's glue, which plays an important role in ensuring the stiffness of structural protein.

TABLE 9.5

Advantages and Disadvantages of Plasma Reactor

Reactor Type	Advantages	Disadvantages
Plasma Reactor	• Production of vitrified, completely inert, non-leachable slag containing heavy metals • The waste products are non-leachable, they can be recycled and used as construction materials right away • Syngas flow rate is reduced • Syngas has a very low content of polluting compounds • Reaction times are extremely fast • There aren't any scaling issues	• Inclusion of nanoparticles in syngas • The presence of moving pieces, as well as their maintenance issues • Refractory energy consumption • Process that is not continuous • Heat shock for the start-up and knock-down • Auxiliary fuel is required to achieve a uniform temperature inside the reactor • Changing the electrodes on a regular basis • Safety problems • In the ducts, molten material solidifies • Expensive plant, operational, and training costs

TABLE 9.6

Gasifiable Biomass

Supply Sector	Type	Example
Forestry	Dedicated forestry	Short rotation plantations (e.g., willow, poplar, eucalyptus)
	Forestry by-products	Wood blocks, wood chips from thinnings
Agriculture	lignocellulosic energy crops that are dry	Herbaceous crops (e.g., miscanthus, reed canary grass, giant reed)
	Oil, sugar and starch energy crops	Oil seeds for methylesters (e.g., rape seed, sunflower)
		Sugar crops for ethanol (e.g., sugar cane, sweet sorghum)
		Starch crops for ethanol (e.g., maize, wheat)
	Agricultural residues	Straw, pruning of vineyards and fruit trees
	Livestock waste	Wet and dry manure
Industry	Industrial residues	Industrial waste wood, sawdust from sawmills
		Fibrous vegetable waste from paper industries
Waste	Dry lignocellulosics	Residues from parks and gardens (e.g., prunings, grass)
	Contaminated waste	Gemolition wood
		Organic fraction of municipal solid waste
		Biodegradable landfilled waste, landfill gas
		Sewage sludge

TABLE 9.7

Composition of Hardwood, Softwood, and Straw (wt % on Dry Basis)

S.No.	Biomass Type	Cellulose	Hemi-Cellulose	Lignin
1.	Hardwood	42–48	27–38	16–25
2.	Softwood	40–45	24–29	26–33
3.	Straws	36–40	21–45	15–20

TABLE 9.8

The Lower Calorific Value of the Main Fuels

Fuel	LHV (MJ/kg)	Bulk Density (Kg/m³)	Energy Density by Volume (MJ/Nm³)
Wood chips (30% MC)	12.5	250	3,100
Logwood (stacked-air dry: 20% MC)	14.7	350–500	5,200–7,400
Wood (solid-oven dry)	19	400–600	7,600–11,400
Wood pellets	17	650	11,000
Miscanthus (bale-25% MC)	13	140–180	1,800–2,300
House Coal	27–31	850	23,000–26,000
Anthracite	33	1,100	36,000
Heating oil	42.5	845	36,000
Natural gas (NTP)	38.1	0.9	35,200
LPG	46.3	510	23,600

A bond made up of cellulose and hemicellulose saccharides forms the long polymeric chains. It is important to have a basic understanding of the physical and chemical properties of the biomass to use it as a sustainable source of energy production.

Biomass varies on the basis of parameters like chemical composition, ashes, inorganic substances, and moisture content (Table 9.7). In the decreasing order of abundance, carbon, oxygen, hydrogen, nitrogen, calcium, potassium, silicon, magnesium, aluminium, sulphur, iron, silicon, chloride, sodium, and manganese are all present (Molino et al., 2014). The type of biomass has little effect on the composition of syngas. It has been observed that using straw as a feedstock increases the hydrogen content while decreasing the lower heating value of the produced gas. When softwood is used as feedstock, the lowest heating value is observed (Sarker and Nielsen, 2015). The amount of char content was highest in woody biomass, while it was lowest in straw biomass. The amount of tar content in hardwood biomass can be found to be higher (Roesch et al., 2011; Kenney et al., 2013). The energy balance of the gasification is largely affected by the moisture content of the biomass; hence, the biomass's moisture content is an important parameter. Table 9.8 exhibits the low heating values of the major fuels.

9.5 Pre-treatment of Biomass

The primary goal of the pre-treatment is to ensure that the material is homogeneous in size and composition, as well as to reduce the initial moisture content of the material, which should not exceed 25–30 wt %. Some of the main pre-treatments other than drying are torrefaction and hydrothermal upgrading.

9.5.1 Torrefaction

Torrefaction occurs in the absence of oxygen at temperatures ranging from 200°C to 300°C. At these temperatures, the biomass loses its rigid fibrous structure and moisture content as the energy content rises. Most of the time, this method is combined with device pelletization to increase the material's bulk density (Van der Stelt et al., 2011). Roasted pellets have an energy per unit volume of 14,000–18,500 MJ/Nm^3, which is substantially higher than pellets made from traditional materials such as wood pellets, which have an energy per unit volume of 8,000–11,000 MJ/Nm^3 (Bobleter, 1994).

9.5.2 Hydrothermal Upgrading (HTU)

This pre-treatment is used to obtain oil, such as crude oil, called bio-crude oil, by decomposing water. It is carried out in two stages. The biomass is treated with water in the first stage at a temperature of 200°C–250°C and a pressure of 30 bar. In the second stage, temperature and pressure rise over a 5- to 10-minute period, allowing biomass conversion to occur by raising the temperature to 300°C–350°C and increasing the 120–180 bar water pressure. The bio-crude oil has a lower heating value in the range of 30–36 M/kg. It can be used as co-fuel in the chemical industry or coal power plant for the production of chemical or synthetic fuel having properties nearly similar to diesel fuel (Dornburg et al., 2006; Peterson et al., 2008).

9.6 Product of Biomass Gasification

The end product of the solid and gas phases differs. Ash, the inert material present in unreacted char and feedstock in the solid phase. There is very little total ash, usually less than 1% by weight. The primary goal is to convert the carbon matrix into gas (Salleh et al., 2010; James et al., 2012). The gas phase, also known as syngas, is a mixture of gas and condensable phases, which consists of CO, H_2, CO_2, CH_4, some C_2-C_3, and light hydrocarbon gases in the gas phase and are incondensable at ambient temperature. If the oxidizing phase contains air as a gasifier carrier, then N_2 is present in this phase as well.

Minor components include NH_3 and inorganic gases such as H_2S and HCL (Lancia et al., 1996; Karatza et al., 1998; Lancia and Musmarra, 1999; Kalisz et al., 2008; Rabou et al., 2009; Yang et al., 2011). The syngas amount can vary from 1 to 3 Nm^3/kg on a dry basis (Wan Ab Karim et al., 2009; Liu et al., 2013). Condensable phases, which are made up of several organic compounds, require special consideration. Tar's composition is determined by feedstock, gasification technology, and operating parameters because it is a condensable hydrocarbon mixture.

The tars based on molecular weight are divided into five classes (Rabou et al., 2009). In class 1 the tar compound is categorized as unknown because of the very high molecular weight and inability to detect by gas chromatography. In class 2, highly water-soluble oxygenated and condensable compounds are included. In classes 3–5, aromatic compound with increasing aromatic ring numbers are included. Single-ring compounds are seen in class 3, while in classes 4 and 5, Polycyclic aromatic hydrocarbon (PAH) compound are included.

9.7 Application of the Gasification Product

The main gasification product is syngas, which is then converted into a final product using proper conversion technology. The primary product obtained from the thermochemical conversion process of biomass can be used for combustion in power plants with conventional fuels like coal-heavy oil and biomass (Kalisz et al., 2008). The higher electrical conversion efficiency can also be obtained by fuel combustion in gas turbines (Bridgwater et al., 2002; Yassin et al., 2009). The combustion of syngas or cracking oil in a boiler is a simple use of the primary thermochemical product in the convectional stem cycle (Yang et al., 2011). Also, the syngas can be used in the internal combustion engine with an overall energy conversion of 35%–45% (Yassin et al., 2009). It can also be used in high-temperature fuel cells. The synthesis can produce methane, and after that, get converted into H_2, which is present in the biomass syngas. The biomethane can also be produced from the syngas by compressing and feeding it into the methanation stage where the exothermal reaction occurs (Habibi et al., 2014).

9.8 Conclusion

Gasification may be a good alternative for energy production from the biomass's energy production, but there is a need to eliminate the causes hindering its development. Further purification is necessary for its best use in the engine. Syngas combustion is more efficient and cleaner than direct biomass combustion. It is a potential alternative to conventional thermochemical processes and intermediate chemical processing, but further research is needed to improve the overall performance of gasification.

REFERENCES

Begum, S., M.G. Rasul, D. Akbar and N. Ramzan. 2013. Performance analysis of an integrated fixed bed gasifier model for different biomass feedstocks. *Energies* 6(12):6508–6524.

Belgiorno, V., G. De Feo, C. Della Rocca and D.R. Napoli. 2003. Energy from gasification of solid wastes. *Waste Management* 23(1):1–15.

Bobleter, O. 1994. Hydrothermal degradation of polymers derived from plants. *Progress in Polymer Science* 19(5):797–841.

Bridgwater, A.V., A.J. Toft, and J.G. Brammer. 2002. A techno-economic comparison of power production by biomass fast pyrolysis with gasification and combustion. *Renewable and Sustainable Energy Reviews* 6(3):181–246.

Canabarro, N., J.F. Soares, C.G. Anchieta, C.S. Kelling and M.A. Mazutti. 2013. Thermochemical processes for biofuels production from biomass. *Sustainable Chemical Processes* 1(1):1–10.

Carpenter, D.L., S.P. Deutchand and R. J. French. 2007. Quantitative measurement of biomass gasifier tars using a molecular-beam mass spectrometer: Comparison with traditional impinger sampling. *Energy & Fuels* 21(5):3036–3043.

Cheng, L., H. Liu, Y. Cui, N. Xue and W. Ding. 2014. Direct conversion of corn cob to formic and acetic acids over nano oxide catalysts. *Journal of Energy Chemistry* 23(1):43–49.

Chhiti, Y., M. Peyrot and S. Salvador. 2013. Soot formation and oxidation during bio-oil gasification: Experiments and modeling. *Journal of Energy Chemistry* 22(5):701–709.

Couhert, C., S. Salvador and J.M. Commandre. 2009. Impact of torrefaction on syngas production from wood. *Fuel* 88(11):2286–2290.

Demirbaş, A., 2001. Biomass resource facilities and biomass conversion processing for fuels and chemicals. *Energy conversion and Management*, 42(11): 1357–1378.

Donatelli, A., P. Iovane and A. Molino. 2010. High energy syngas production by waste tyres steam gasification in a rotary kiln pilot plant. Experimental and numerical investigations. *Fuel* 89(10):2721–2728.

Dornburg, V., A.P. Faaij and B. Meuleman 2006. Optimising waste treatment systems: Part A: Methodology and technological data for optimising energy production and economic performance. *Resources, Conservation and Recycling* 49(1):68–88.

Gómez-Barea, A., R. Arjona and P. Ollero. 2005. Pilot-plant gasification of olive stone: A technical assessment. *Energy & Fuels* 19(2):598–605.

Habibi, N., M. Rezaei, N. Majidian and M. Andache. 2014. CH_4 reforming with CO_2 for syngas production over La_2O_3 promoted Ni catalysts supported on mesoporous nanostructured γ-Al_2O_3. *Journal of Energy Chemistry* 23(4):435–442.

Hamelinck, C.N., A. P. Faaij, H. den Uil and H. Boerrigter. 2004. Production of FT transportation fuels from biomass; technical options, process analysis and optimization, and development potential. *Energy* 29(11):1743–1771.

Higman, C. and M. van der Burg. 2008. Chapter 2-the thermodynamics of gasification. Gasification. Burlington: Gulf Professional Publishing, 28–9.

Iovane, P., A. Donatelli, and A. Molino. 2013. Influence of feeding ratio on steam gasification of palm shells in a rotary kiln pilot plant. Experimental and numerical investigations. *Biomass and Bioenergy* 56:423–431.

James, A.K., R.W. Thring, S. Helle and H.S. Ghuman 2012. Ash management review—applications of biomass bottom ash. *Energies* 5(10):3856–3873.

Jansen, R.A., 2012. *Second Generation Biofuels and Biomass: Essential Guide for Investors, Scientists and Decision Makers*. John Wiley & Sons, Hoboken, NJ.

Kalisz, S., M. Pronobis and D. Baxter. 2008. Co-firing of biomass waste-derived syngas in coal power boiler. *Energy* 33(12):1770–1778.

Karatza, D., A. Lancia and D. Musmarra, 1998. Fly ash capture of mercuric chloride vapors from exhaust combustion gas. *Environmental Science & Technology* 32(24):3999–4004.

Kenney, K.L., W.A. Smith, G. L. Gresham and T. L. Westover. 2013. Understanding biomass feedstock variability. *Biofuels* 4(1):111–127.

Lancia, A. and D. Musmarra. 1999. Calcium bisulfite oxidation rate in the wet limestone– gypsum flue gas desulfurization process. *Environmental Science & Technology* 33(11):1931–1935.

Lancia, A., D. Karatza, D. Musmarra and F. Pepe. 1996. Adsorption of mercuric chloride from simulated incinerator exhaust gas by means of Sorbalit TM particles. *Journal of Chemical Engineering of Japan* 29(6):939–946.

Li, S., S. Xu, S. Liu, C. Yang and Q. Lu. 2004a. Fast pyrolysis of biomass in free-fall reactor for hydrogen-rich gas. *Fuel Processing Technology* 85(8–10):1201–11.

Li, X. T., J. R. Grace, C. J. Lim, A.P. Watkinson, H. P. Chen and J. R. Kim. 2004b. Biomass gasification in a circulating fluidized bed. *Biomass Bioenergy* 26(2):171–93.

Li, X.T., J. R. Grace, C. J. Lim, A. P. Watkinson, H. P. Chen and J. R. Kim. 2004. Biomass gasification in a circulating fluidized bed. *Biomass Bioenergy* 26:171–193.

Liu, B. and S. Ji. 2013. Comparative study of fluidized-bed and fixed-bed reactor for syngas methanation over Ni-W/TiO$_2$-SiO$_2$ catalyst. *Journal of Energy Chemistry* 22(5):740–746.

Liu, Q., L. Liao, Z. Liu and X. Dong. 2013. Hydrogen production by glycerol reforming in supercritical water over Ni/MgO-ZrO$_2$ catalyst. *Journal of Energy Chemistry* 22(4):665–670.

Lv, P.M., Z. H. Xiong, J. Chang, C. Z. Wu, Y. Chen and J. X. Zhu. 2004. An experimental study on biomass air–steam gasification in a fluidized bed. *Bioresource Technology* 95(1):95–101.

Mc. Kendry, P. 2002. Energy production from biomass (part 1): Overview of biomass. *Bioresource Technology* 83(1):37–46.

Mohammed, M. A. A., A. Salmiaton, W. A. K. G. Wan Azlina, M. S. Mohammad Amran and A. Fakhru'l-Razi. 2011. Air gasification of empty fruit bunch for hydrogen-rich gas production in a fluidized-bed reactor. *Energy Conversion and Management* 52(2):1555–61.

Molino, A., F. Nanna and A. Villone. 2014. Characterization of biomasses in the southern Italy regions for their use in thermal processes. *Applied Energy* 131:180–188.

Molino, A., P. Iovane, A. Donatelli, G. Braccio, S. Chianese and D. Musmarra. 2013. Steam gasification of refuse-derived fuel in a rotary kiln pilot plant: Experimental tests. Chemical Engineering Transactions 32.

Molino, A., S. Chianese and D. Musmarra. 2016. Biomass gasification technology: The state of the art overview. *Journal of Energy Chemistry* 25(1):10–25.15

Paula, A., G. Peres, B.H. Lunelli and R. M. Fllho. 2013. Application of biomass to hydrogen and syngas production. *The Italian Association of Chemical Engineering* 32:589–94.

Peterson, A.A., F. Vogel, R.P. Lachance, M. Fröling, M.J. Antal Jr. and J.W. Tester. 2008. Thermochemical biofuel production in hydrothermal media: A review of sub-and supercritical water technologies. *Energy & Environmental Science* 1(1):32–65.

Qian, K., A. Kumar, K. Patil, D. Bellmer, D. Wang, W. Yuan and R. L. Huhnke. 2013. Effects of biomass feedstocks and gasification conditions on the physiochemical properties of char. *Energies* 6(8):3972–3986.

Rabou, L.P., R.W. Zwart, B. J. Vreugdenhil, and L. Bos. 2009. Tar in biomass producer gas, the Energy research Centre of the Netherlands (ECN) experience: An enduring challenge. *Energy & Fuels* 23(12):6189–6198.

Roesch, H., J. Dascomb, B. Greska and A. Krothapalli. 2011. Prediction of producer gas composition for small scale commercial downdraft gasifiers. In *Proceedings of 19th European Biomass Conference and Exhibition*, Berlin, Germany (pp. 1594–1601).

Roos, C.J. 2010. Clean heat and power using biomass gasification for industrial and agricultural projects. Northwest CHP Application Center.

Salleh, M., N.H. Kisiki, H. M. Yusuf and W. A. Ab Karim Ghani. 2010. Gasification of biochar from empty fruit bunch in a fluidized bed reactor. *Energies* 3(7):1344–1352.

Sarker, S. and H. K. Nielsen. 2015. Assessing the gasification potential of five woodchips species by employing a lab-scale fixed-bed downdraft reactor. *Energy Conversion and Management* 103:801–813.

Schmid, J.C., U. Wolfesberger, S. Koppatz, C. Pfeifer and H. Hofbauer. 2012. Variation of feedstock in a dual fluidized bed steam gasifier—influence on product gas, tar content, and composition. *Environmental Progress & Sustainable Energy* 31(2):205–215.

Shen, D., R. Xiao, S. Gu and K. Luo. 2011. The pyrolytic behavior of cellulose in lignocellulosic biomass: A review. *RSC Advances* 1: 1641–1660.

Van der Drift, A., H. Boerrigter, B. Coda, M.K. Cieplik and K. Hemmes. 2004. Entrained flow gasification of biomass. Ash behaviour, feeding issues, system analyses.

Van der Stelt, M. J. C., H. Gerhauser, J.H.A. Kiel and K. J. Ptasinski. 2011. Biomass upgrading by torrefaction for the production of biofuels: A review. *Biomass and Bioenergy* 35(9):3748–3762.

Wan Ab Karim Ghani, W.A., R.A. Moghadam, A. Salleh and A.B. Alias. 2009. Air gasification of agricultural waste in a fluidized bed gasifier: hydrogen production performance. *Energies* 2(2):258–268.

Widyawati, M., T.L. Church, N.H. Florin and A. T. Harris. 2011. Hydrogen synthesis from biomass pyrolysis with in situ carbon dioxide capture using calcium oxide. *International Journal of Hydrogen Energy* 36(8):4800–4813.

Wu, Y., W. Yang and W. Blasiak. 2014. Energy and exergy analysis of high temperature agent gasification of biomass. *Energies* 7(4):2107–2122.

Yang, W., D.J. Yang, S.Y. Choi and J.S. Kim. 2011. Experimental study on co-firing of syngas as a reburn/alternative fuel in a commercial water-tube boiler and a pilot-scale vertical furnace. *Energy & Fuels* 25(6):2460–2468.

Yassin, L., P. Lettieri, S.J. Simons and A. Germanà. 2009. Techno-economic performance of energy-from-waste fluidized bed combustion and gasification processes in the UK context. *Chemical Engineering Journal* 146(3):315–327.

Zhang, Q., L. Dor, L. Zhang, W. Yang and W. Blasiak. 2012. Performance analysis of municipal solid waste gasification with steam in a Plasma Gasification Melting reactor. *Applied Energy* 98:219–229.

10

Solar Energy Challenges, Gadgets, and Their Application

Shambhavi Yadav, Anugrah Tripathi, and Garima Kumari

CONTENTS

10.1 Introduction

With the increasing demand of energy via greener methods and the ongoing exhaustion of fossil fuels, solar energy conversion has regained the spotlight of the global energy activities. Scientists in the last 4–6 decade have explored the opportunity of using alternative energy sources such as solar, wind, and biomass for various purposes. Solar energy can be converted in many forms and used in different form of energy, *e.g.* solar drying, solar water heating system, solar furnace, solar lighting, etc. As far as renewable energy sources are concerned, solar thermal energy is the most abundant one and is available in both direct as well as indirect forms. Since demand for energy is rising each year and will continue so for years to come, capacity addition is of paramount importance to reduce energy crisis and power deficits as well as control greenhouse gas (GHG) emissions as per globally accepted norms. Most of the developing countries still rely on fossil fuels to meet their energy demands and is a major concern for global climate change. In India, which is one of the fastest developing countries, a major percent of total power is generated through thermal power plants and burning of fossil fuels for electricity generation, which alone contributes heavily to total GHGs released. Nevertheless, for development of a country, power generation is a necessity, but more the installed capacity of power plants utilizing fossil fuels, higher are the risks of climate change and environment pollution. To create a balance, use of alternative forms of energy,

DOI: 10.1201/9781003175926-10

like solar and wind, becomes imperative. These forms of energy are renewable, available in surplus, are environment friendly, and power generation using solar- or wind-based plants produce minimal levels of pollution whatsoever (Singal, 2007). For sustainable growth and development, countries need to invest in these forms of energy and explore the benefits and wider utilization of solar-based power generation systems to achieve and realize the targets of highly anticipated global climate change goals (Fthenakis et al., 2009).

Realizing the benefits of solar energy, there have been continuous developments in the technology, manufacture, and installation of solar-based products and facilities around the world. Several new gadgets have been introduced into market that simplify the living, have potential to uplift livelihoods, and are user- as well as environment-friendly. According to latest reports by International Renewable Energy Agency (IRENA), the total installed solar photovoltaic systems worldwide has reached over 580 GW by the end of year 2019 (IRENA, 2020), which represents a 14-fold growth since the year 2010. This substantial increase in growth has been mainly due to addition of new capacities in the Asian region. The Asian region alone contributed to 60% of growth recorded for the year 2019 with countries like India, Japan, China, and the Republic of Korea together, installing 47.5 GW of new photovoltaic capacities (IRENA, 2020). Similar to developments in photovoltaics, the cumulative concentrated solar power has also grown significantly and reached about 6.3 GW till 2019.

With several benefits and advantages, solar power has immense potential to be utilized as a future powerful energy source. However, there are many challenges as well that are faced at different levels of solar technology development and implementation. The present chapter also discusses the challenges of utilizing solar power at environmental, financial, technological, as well as policy levels. In brief, there are difficulties when it comes to allocation of lands for solar power plants, and complicated government guidelines and regulations further hinder the process and may suppress interests of people as well as industries. Then there are technological challenges in the form of advancement of research in the field, setting up manufacturing units and several others. Continuous flow of funds and financial subsidies by governments is required to keep solar energy sector flourishing. Environmental issues are much less when compared to conventional fossil fuel-based energies but pollution caused during manufacturing of semiconductor devices, cells, and panels required in various solar-based products is needed to be kept under control and regular monitoring. Also, any disturbances to the local flora and fauna are one of the important issues to be handled carefully while setting up any solar production unit.

10.2 Need to Harness Solar Power

The increasing world population has undoubtedly led to emergence of numerous industries all across developed and developing countries, thereby increasing our dependency on fossil fuels as source of energy. This increased usage of fossil fuels has created a risk of exhausting the non-renewable resources and ignited scientific interests in exploring alternative sources of energy to meet the growing demands (Bilgen et al., 2004; Demirbas, 2006). Among the available renewable energy sources, solar energy harnessed by the sun is abundantly utilized in various applications. There are many benefits of using solar energy as it offers a clean energy alternative and can be used in remotest of places around the world (Goswami et al., 2004). People have been harnessing solar energy from ancient times for drying of agricultural produce, and its use in domestic solar cooking devices has also been well acknowledged. The immense potential available with solar energy, if harnessed properly, could meet the energy demands of various industries, provide new options of livelihood upliftment in remote areas, and overall help us meet the global goals of sustainability (Okoro, 2006). Environmental pollution, one of the biggest negative impacts of fossil fuel usage, can be controlled to a larger extent by promoting solar power-based energy generation programs (Dincer, 2001).

Although there are significant challenges associated with adopting solar energy, there are fair shares of incentives that could positively affect the growth and development of solar technology in a country. First, availability of a significant amount of annual average direct normal irradiance in a country is one of the major factors that drive interest towards solar power generation. So, ideally, the aim should be to

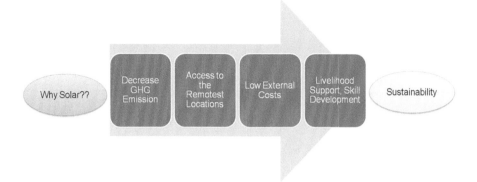

FIGURE 10.1 Benefits of harnessing solar power. Use of solar energy diminishes emission of greenhouse gases and hence provides clean energy as compared to fossil fuels. Solar energy-based power generation has low external costs and even remotest of locations can benefit from these. It can lead to livelihood upliftment and skill development in such areas.

recognize and select solar hotspot areas and promote them for solar energy-based programs. Second, it has been projected that externalities of a solar power project, especially in case of utility scale solar energy-based power plants, are much lower than other power projects based on fossil fuels. For example, the amount of pollution out of solar power plants is negligible. Also, there is no risk of fuel shortage or extreme utilization of water resources. All these factors add on to the benefits of solar energy in power generation. Lastly, for any country or region to start a power generation project, requirement of land is one of the most important prerequisites. In case of solar energy, wastelands can also be utilized for this purpose, which also reduces environmental impact by not disturbing cultivable lands or habitats (Rathore et al., 2017). Figure 10.1 highlights major benefits of utilizing solar power that ultimately lead to sustainable development.

10.3 Challenges Associated with Use of Solar Power

Although solar power promises several benefits and is one of the most potential renewable energy resources, there are plenty of challenges associated with its use, development of technology and products, as well as promotion and applicability. Figure 10.2 depicts some of the main challenges associated with utilization and application of solar energy-based systems. It is a challenge to apply solar powered devices in areas that receive very less sunlight, especially the colder and snow-prone areas of the world. Also, variances in the intensity of sunlight may affect the efficiency and energy generation through solar cells. Nevertheless, although there are immense benefits associated with utilization of solar power, setting up a solar power plant or any large-scale solar-based energy generation facility requires vast area of land, which is a major challenge in recent scenarios. This may sometimes lead to agricultural lands being taken up and converted into power stations. Also, the local flora and fauna may get disturbed due to any such activity, which should also be taken care of while choosing a location for establishment of solar plants or units.

The photovoltaic (PV) cells used to convert solar energy into electric energy are a key component of solar power system, and enhancing the efficiency of PV cells is a challenge and continuous research is going on in this area (Sherwani et al., 2010). Efficient PV cells would ensure maximum conversion of available light energy into useful form and would be more cost-effective. Hayat et al. (2019) discussed in detail the various challenges associated with solar energy use and emphasized on development of cost-effective, less expensive, more efficient PV cells to ensure maximum energy production, storage, and availability. Also, there are challenges associated with the availability of raw materials that are required

FIGURE 10.2 Challenges associated with solar power. As discussed in detail in the text, various factors affect the successful implementation of a solar power program and these can be financial, technological, geographical, environmental, policy and regulation, manufacturing, transmission and distribution, etc.

during manufacturing of solar cells as well as cost of their production, which ultimately affects the economics and cost of electricity generation through solar power. Although in recent years, costs are reduced due to advancements in technology, according to latest reports by International Renewable Energy Agency (IRENA), renewable energy costs are increasingly becoming cost-effective. Costs for electricity produced from solar-based photovoltaics (PV) fell 82% in the last decade from 2010 and 2019 (IRENA, 2020).

The impact of solar technologies is evaluated on the economic, environmental, as well as social levels. For countries where research and development is not so strong, there are many technological challenges in implementing solar power systems. Economics thus becomes a major driver as huge amount of initial money is invested in research, manufacture, and efficient development of solar-based devices, especially PVs. In such cases, ensuring the benefits of solar power to local and less privileged communities becomes difficult since costs become higher. Boosting manufacturing at domestic levels and skill development of local people can provide additional benefits and cost cutting. Researchers have suggested implementation of decentralized off-grid PV systems that are financially and logistically more beneficial and could be utilized well for most vulnerable localities (Mandelli et al., 2016). Also, regulatory policies framed by governments play a key role in development and progress of solar projects. Challenges exist in policy framework wherein land acquisition is a major factor (Rathore et al., 2017). Also, smooth transmission and distribution is required for efficient utilization of generated power, and government support is paramount in such cases. Creation of a conducive environment will certainly boost the renewable energy sector, including the solar energy sector.

10.4 Solar Gadgets and Applications

In this section, various gadgets that have been developed to harness and use solar energy in different applications are discussed. Starting from cooking devices to advanced PV systems, advancements have been made at technological levels to deliver high-performance, cost-effective, and environment-friendly solar appliances for domestic as well as industrial use.

10.4.1 Solar Panels or Solar Collectors

Solar energy collectors are specially designed heat exchangers that are used to transform energy from solar radiation into internal energy of the transport medium. These are one of the major components of any solar power system. Two types of solar collectors are described: non-concentrating or the stationary collectors and the concentrating collectors. Interception and absorbance of solar radiations is done in same area in case of non-concentrating collectors, whereas reflecting concave surfaces are used to track and intercept solar radiations in concentrating collectors that then get absorbed through a smaller area (Aghaei, 2014; Hayat et al., 2019).

10.4.2 Solar Heaters

Solar energy can also be used for heating purposes, and devices like solar water heaters and solar air heaters have been designed and are available on the market. Solar water heaters are very successful for large rooftops over high-rise buildings that receive uninterrupted sunlight throughout the day. Solar water heating devices comprise of solar panels that absorb and concentrate sunrays and generate heat energy, which is then transferred to the circulating water inside the metal tubing attached to the system. Hot water gets stored in insulated tanks for further use. The circulation of water can be achieved by either natural or forced circulation methods. Both small- as well as large-scale solar water heaters are well in use by various consumers (Aghaei, 2014; Sudheer et al., 2018). Similarly, air heating can also be done using solar panels that are connected to air ducts and circulation of warm air in buildings can be achieved. Sometimes, the heat generated is also used to heat circulating fluids used in household heaters during cold season.

10.4.3 Solar Cookers

Use of solar energy in cooking has been one of the direct applications of this renewable source. Solar cooking devices are based on the principle that sunlight is first concentrated using reflectors or mirrors and focused on to a surface preferably black or of a designated material that absorbs maximum sunlight and converts it into heat energy. Additionally provisions are made to trap this heat energy by insulating the devices, thereby increasing the efficiency. There are two basic types of solar cookers that are prominently used: the Box type and the Concentrator type, the former being widely used in domestic cooking and later mainly used for large-scale community purposes.

In the box type solar cooker, the cooking box or tray is of black metal or is painted black and the cooking vessel such as pots or pans are kept inside this black tray. The tray is enclosed within another box with insulation in between the two layers of boxes to avoid heat loss. A glass sheet, usually double glass, is used as lid over the metallic box, and this allows the incoming solar radiations to enter inside and heat up the metallic surface as well as the cooking vessel. The glass lid almost creates a greenhouse effect by minimally allowing any emitted radiation to leave the box. This increases temperature inside the cooker that is then absorbed by the food inside the vessel and it gets cooked. Additionally, mirrors are also placed to focus sunrays on to the glass aperture (Sudheer et al., 2018). However, solar cookers, whether domestic or industrial, have not been able to be much popular among masses due to limitations such as inefficient working on a cloudy day, inability to function at nights, higher costs, and more time required for cooking.

10.4.4 Solar Dryers

Drying of agricultural produce by exposing it to sunlight has been practiced since ancient times. With the advancement of solar energy technology, devices such as solar dryers have been developed that are based on similar principles as solar cookers. Sunrays are trapped into an enclosed area using reflective mirrors and glasses. This heats up the interior of the dryer where the crop products are kept and drying is carried out. Dryers have been designed with facilities of natural and convection type forced air circulation features that make the drying process faster. Additionally, removal of vapours, generated from crop produce drying, can also be provided in the solar dryer that increases its utility. Separate air inlet and exhaust valves are incorporated for these purposes (Kumar et al., 2016).

10.4.5 Photovoltaic cells or Solar Cells

Solar cells based on the PV technology are the base of many solar-based gadgets and products available in market, viz. solar watches, solar calculators, solar textiles, and many more. A solar cell directly converts the solar energy into electrical energy (DC). Semiconductors are the most important components of the solar cells, and mainly silicon is widely used. The electricity generated can also be stored in batteries and used in invertors to supply electric current in remote locations. Products such as solar street lights and solar lights or lanterns are widely used in various countries, especially in areas where electricity supply faces frequent interruptions. These use solar PV panels which absorb sunlight and convert it to electric current.

10.4.6 Solar Vehicles

It is well known that fossil fuels have a major role to play in the progress of automobile industry. Nevertheless, with air pollution becoming a major problem in most of the big cities and global climate change seen as the biggest threat to mankind's survival, newer and cleaner technologies are being sought. Solar-powered vehicles, although still in the research and development pipeline, are coming up as a promising alternative to fossil fuel-based vehicles. Many automobile giants and start-ups are working towards developing solar cars, and the technology could well change the future of the automobile industry. Solar vehicles also rely on the PV technology to convert solar energy into electricity. Energy is also stored in batteries that allow vehicle function even in absence of sunlight. Several prototypes of solar-powered vehicles, especially solar cars, are currently under development by both large as well as small automakers. According to some estimates, the solar vehicle market could reach more than $600 billion by 2027. Innovations in design are coming up that could allow maximum possible solar panel over the vehicle without compromising design and aesthetics.

10.5 Solar Energy and Sustainability

Major electricity production methods are based on coal, oil, and natural gas, all of which contribute to global climate change in one way or the other. In this regard, renewable energy such as solar energy can play a crucial role in bringing down global temperatures and ensure sustainable development. In India alone, about 74% of energy demands are met through oil and coal, which itself indicates to the amount of carbon emissions released during the process (Kumar and Majid, 2020). The Indian Government has targeted the installation of about 175 GW of renewable power capacity by the year 2022, increasing its share in non-fossil fuel-based electricity capacity to 40% by year 2030 (Annual Report 2019–2020, Ministry of New and Renewable Energy, Government of India). Although solar energy is considered as a clean form of energy that does not causes air or noise pollution while in use, there have been concerns over the amount of pollution caused or threat posed to the environment during manufacturing of solar cells and other components. If we consider the process of battery manufacture, it requires several minerals that are to be mined from Earth and hence contribute to carbon foot print. Here, we discuss possible environmental benefits and implications of solar energy utilization and generation. Although the upfront costs of solar PV panels are usually high, the long-term gains in the form of zero emissions make the technology promising and highly desirable for future. Solar energy has potential to reduce or limit carbon emissions in major areas such as transportation, industry, electricity generation, commercial sectors, as well as domestic residential areas. Hernandez et al. (2014), while reviewing the impacts of utility-scale solar installations in rural areas, found that these have very low environmental impacts when compared to other forms of energy, especially fossil fuels. Decrease in GHG emission in cities can also be achieved through implementation of solar roof-top panels that have proved to be cost-effective as well as useful in renewable energy generation (Arnette, 2013).

10.6 Solar Energy and Agriculture

Solar energy can be used in all aspects of agriculture. This will undoubtedly aid in meeting the growing demand for agricultural products as the world's population grows. Much farm energy needs can be met or supplemented with solar energy. One of the oldest and most commonly used applications of solar energy is drying crops and grains in the sun. Allowing crops to dry naturally in the field or spreading grain and fruit out in the sun after harvesting are the easiest and least costly methods. These methods have the downside of exposing crops and grains to harm from birds, rodents, wind, and rain, as well as pollution from wind-blown dust and soil.

Advanced solar dryers dry faster, protect grain and fruit thereby reduce losses, and provide a higher-quality product than open-air methods (EREC, 2003; UCS, 2009). Other than this, commercial greenhouses normally use the sun for illumination, but they are not built to use the sun for heating. During the colder months, they rely on gas or oil heaters to sustain the temperatures needed for plant growth. Solar greenhouses, on the other hand, are designed to use solar energy for both heating and lighting (Chikaire et al., 2010; Kumar et al., 2019).

In areas where there is no existing power line, PV water pumping systems may be the most cost-effective water pumping option. They are ideal for grazing operations that need to supply water to far-flung pastures. Simple PV power systems run pumps directly when the sun shines, so they are at their most efficient during the hot summer months when they are most needed (EREC, 2003; NYSERDA, 2009).

10.7 Solar Energy and Forest

Our forest is an outcome of the use of solar radiation. Insolation is a phenomenon for combining elements located near the earth's surface in order to complete critical physiological processes for forest growth and maturity. The hydrological cycle in our forested watersheds is also driven by the light. A case study of Yunan in the year 2015 is the best example providing strong evidence that advanced PV materials will protect forests from forest fires and other threats. They designed a monitoring network powered by an off-grid solar power system with advanced PV materials.

10.8 Solar Energy and Environment

The greatest benefit of solar energy systems is that they are environmentally friendly. Solar energy systems outperform fossil fuel-powered systems by a wide margin. It has been discovered that domestic solar water heating systems emit 80% less greenhouse gases than traditional systems. When compared to a traditional electric power plant system, solar space heating reduces greenhouse gas emissions by 40%. Solar energy has many environmental (lower carbon dioxide emissions) and socioeconomic benefits. By switching from conventional energy sources to solar energy technologies, we can prevent our water resources from pollution and can prevent our air from emitting noxious gases like CO_2, SO_2, etc. (Tsoutsos et al., 2005).

10.9 Solar Energy and Climate Change

Climate change is one of the primary concerns for humanity in the 21st century. Global climate change may be the most significant environmental issue relating to fossil fuel (global warming or the greenhouse effect). The steady rise in harmful gas emissions due to the use of fossil fuels has resulted in a strong demand for renewable energy as a base load, such as solar energy (Tsoutsos et al., 2005). The growing concentration of greenhouse gases in the atmosphere, such as CO_2, CH_4, CFCs, halons, N_2O, ozone, and peroxyacetyl nitrate, is trapping heat radiated from Earth's surface and raising the surface temperature (Dincer, 2001). The majority of farm machinery is driven by fossil fuels, which lead to greenhouse gas emissions and, as a result, accelerate climate change.

10.10 Conclusion

Given that fossil fuels will run out one day, finding alternative fuels will be necessary. Renewable energy is seen as a viable alternative to fossil fuels, and it is receiving a lot of publicity these days. The Sun is an endless source of energy that can meet all energy requirements of humanity. Solar is the most effective renewable energy source since it is available in every part of the world. Solar energy has the potential to be the most viable source of energy in the future. Solar-powered electricity generation appears to be very promising. In addition to generating electricity, solar energy has the potential to meet the energy needs of important working sectors of society, such as agriculture, water treatment, etc. Solar energy production is expected to be able to compete economically with traditional energy sources by 2032, according to estimates. Solar energy is also the most environment-friendly energy source available. Unlike other fossil fuels, this fuel does not pollute the environment, mitigates CO_2 emission, and has the potential for safe and sustainable development.

REFERENCES

Aghaei, T.P. 2014. Solar Electric and Solar Thermal Energy: A Summary of Current Technologies. Global Energy Network Institute (GENI).

Annual Report 2019–2020. Ministry of New and Renewable Energy, Government of India.

Arnette, A. N. 2013. Integrating roof top solar into a multi-source energy planning optimization model. *Applied Energy*. 111:456–467.

Bilgen, S., K. Kaygusuz and A. Sari. 2004. Renewable energy for a clean and sustainable future. *Energy Sources, Part A: Recovery, Utilization, and Environmental Effects* 26(12):1119–29.

Chikaire, J., F. N. R. N. Nnadi, N. O. Nwakwasi, O. O. Anyoha, P. A. Aja Onoh and C. A. Nwachukwu. 2010. Solar energy applications for agriculture. *Journal of Agricultural and Veterinary Sciences* 2: 58–62.

Demirbas, A. 2006. Global renewable energy resources. *Energy Sources, Part A: Recovery, Utilization, and Environmental Effects*. 28 (8):779–92.

Dincer, I. 2001. Environmental issues: I-energy utilization. *Energy Sources, Part A: Recovery, Utilization, and Environmental Effects*. 23 (1):69–81.

EREC. 2003. Agricultural Applications of Solar Energy. Energy Efficiency and Renewable Energy Cleaning house (EREC) United State Department of Energy, Merrifield. Available at www.p2pays.org/ref/24/23989.htm.

Fthenakis, V., J.E. Mason and K. Zweibel 2009. The technical, geographical, and economic feasibility of solar energy to supply the energy needs of the US. *Energy Policy* 37:387–99.

Goswami, D.Y., S. Vijayaraghavan, S. Lu and G. Tamm. 2004. New and emerging developments in solar energy. *Solar Energy* 76:33–43.

Hayat, M.B., D. Ali, K.C. Monyake, L. Alagha and N. Ahmed, 2019. Solar energy-A look into power generation, challenges, and a solar-powered future. *International Journal of Energy Research* 43(3):1049–1067.

Hernandez, R.R., S.B. Easter, M. L. Murphy-Mariscal, F. T. Maestre, M. Tavassoli, E. B. Allen, C. W. Barrows, J. Belnap, R. Ochoa-Hueso, S. Ravi and M. F. Allen. 2014. Environmental impacts of utility-scale solar energy. *Renewable and Sustainable Energy Reviews* 29:66–779.

IRENA. 2020. Renewable Power Generation Costs in 2019, International Renewable Energy Agency, Abu Dhabi.

Kumar, N., N. Jeena, and H. Singh 2019. Elevated temperature modulates rice pollen structure: A study from foothill Himalayan Agro-ecosystem in India. *3Biotech* 9: 175.

Kumar, M., S.K. Sansaniwal and P. Khatak. 2016. Progress in solar dryers for drying various commodities. *Renewable and Sustainable Energy Reviews* 55:346–360.

Kumar, J.C.R. and M.A. Majid. 2020. Renewable energy for sustainable development in India: current status, future prospects, challenges, employment, and investment opportunities. *Energy, Sustainability and Society* 10(2): 1–36.

Mandelli, S., J. Barbieri, R. Mereu and E. Colombo. 2016. Off-grid systems for rural electrification in developing countries: Definitions, classification and a comprehensive literature review. *Renewable and Sustainable Energy Reviews* 58:1621–1646.

NYSERDA. 2009. Introduction to Solar Energy Applications for Agriculture. New York State Energy Research Development Authority, New York. Available at www.power Naturally.org.

Okoro, O.I. and T. C. Madueme. 2006. Solar energy: A necessary investment in a developing economy. *International Journal of Sustainable Energy* 25(1):23–31.

Rathore, P.K.S., S. Rathore, R.P. Singha and S. Agnihotri. 2017. Solar power utility sector in India: Challenges and opportunities. *Renewable and Sustainable Energy Reviews*. http://dx.doi.org/10.1016/j.rser.2017.06.077.

Sherwani, A.F., J.A. Usmani and Varun. 2010. Life cycle assessment of solar PV based electricity generation systems: A review. Renewable and Sustainable Energy Reviews 14:540–544.

Singal, S.K. 2007. Review of augmentation of energy needs using renewable energy sources in India. *Renewable and Sustainable Energy Reviews* 11(7):1607–1615.

Sudheer, K.P, P.K. Sureshkumar, S. V. Sreekutty and K. Greeshma. 2018. Lecture Notes of Renewable Energy. Published by The Associate Dean, College of Horticulture, Vellanikkara, Kerala Agricultural University.

Tsoutsos, T., F. Niki and G. Vassilis. 2005. Environmental impacts from the solar energy technologies. *Energy Policy* 33(3): 289–296.

UCS. 2009. Renewable Energy and Agriculture: A Natural Fit. Union of Concerned Scientists, Cambridge. Available at www.ucssusa.org/elean-energy/coalvswind/gd.

11

Impact of Renewable Energy in Reducing Greenhouse Gas Emissions

Megha Verma, Meenakshi Sati, and Shivani Uniyal

CONTENTS

11.1 Introduction

In the past few decades, it has become clear that human activity influences the climate of Earth. Between 1910 and 2009, the global surface temperature increased at a rate of 0.7°C–0.75°C every 100 years (Wen et al., 2011). Activities such as fuel combustion change in land use patterns and agricultural production increase the concentration of greenhouse gases and aerosols in the atmosphere. Carbon dioxide (CO_2) is the main GHG released by fuel combustion, accounting for 76% of overall GHG emissions (Fay and Golomb, 2002). In recent years, climate change has become the focus of global attention (Pielke et al., 2007; Hotchkiss et al., 2015). The significant impact of climate change is rising the world's temperature, decreasing the snow cover, and increasing the sea level. No doubt that the temperature increase due to climate change will interact with the hydrological cycle and result in the change in the pattern of precipitation, evapotranspiration, and soil moisture, melting glacier ice and ice caps and causing river flow variability. These changes will have impacts on water resources and water supply, floods and droughts, and hydropower generation (Stocker, 2013). Serious actions are taken to reduce global warming and greenhouse gas emissions. The world could be heading not only towards the reduction of global warming but also, and more importantly, towards a major environmental disaster (Menyah and Wolde-Rufael, 2010; Reddy and Assenza, 2009). Despite many efforts, greenhouse gas emissions have risen very fast (Friedlingstein et al., 2019; Connolly et al., 2020). After the Industrial Revolution, most of the rise in greenhouse gas emissions (chiefly carbon dioxide, CO_2) since the 19th century is due to the use of fossil fuel-generated energy (coal, oil, natural gas, and peat) (Liu et al., 2011).

DOI: 10.1201/9781003175926-11

Due to the long lifetime of CO_2 (estimated in the the100–300 year range) even if we stop combusting fossil fuels and discharging CO_2 into the atmosphere, the average global temperature of the Earth will continue to increase for the rest of the century, meaning that the excess atmospheric stocks (515Gt Carbon) would continue to drive radiative forcing and global warming for many decades (Matthews and Caldeira, 2008). Second, even if atmospheric concentrations were to decrease, CO_2 would outgas from the oceans and offset this decrease because of the dynamic equilibrium between the CO_2 in the atmosphere and the (bi)carbonates (HCO_3^-/CO_3^{2-}) dissolved in the oceans (Vichi et al., 2013). Third, besides CO_2, about 34% of radiative forcing is contributed by other GHGs (Tian et al., 2016). Even if all excess anthropogenic atmospheric CO_2 were removed, radiative forcing would only be reduced by half (Cao and Caldeira, 2010).

Greenhouse gas (GHG) emissions due to human activity have been altering our planet's energy and climatic patterns. These emissions increased the atmospheric concentration of CO_2 from ~277 ppm in 1750 to 397 ppm in 2014 increment of about 43%. NASA reported that the hottest year was 2015. In 2015, CO_2 levels registered several peaks over 400 ppm in March and December (Berga, 2016). Efforts to use renewable energy sources are driven primarily by greenhouse gas emissions (Banos et al., 2011). In the World Climate Conference held in Paris, 196 parties (including 195 separate countries and the entire European Union) came to an agreement to reduce global temperature rises to no more than 2°C. According to a previously published report by the Intergovernmental Panel on Climate Change (IPCC) (O'Neill et al., 2017), CO_2 emissions have historically always increased in line with improvements in the global economy. However, the United States is currently experiencing negative growth in CO_2 emissions, while the country's GDP per capita has continued to grow since 2007. According to BP Statistical Review of World Energy 2016 (Wang and Li, 2016; Sakata et al., 2017; Lee et al., 2017), US carbon emissions increased year-on-year before 2007. One of the significant achievements by the United Nations and the partner countries was the approval of the Paris Agreement on Climate Change on December 12, 2015, at the United Nations Framework Convention on Climate Change (UNFCCC, 2015). This convention aims to reduce the emission of greenhouse gas (GHG) such that their concentrations in the atmosphere stop increasing, and this is a very significant global challenge. To achieve this goal, it is required to electrify transportation, generate electricity without carbon emissions, electrify agricultural operations, heat buildings with solar energy and electricity, and reduce carbon emissions associated with construction, mining, and industrial production (Williams et al., 2012).

Pielke (2005) noted that there are two approaches to reduce the impacts of future climate change: (i) "climate mitigation" and (ii) "climate adaptation. The first approach, "climate mitigation", assumes that greenhouse gases are the primary driver of climate change and tries to "reduce future climate change" by reducing greenhouse gas emissions. The second approach, "climate adaptation", involves developing better systems and infrastructure for dealing with climate change and climate extremes. Pielke Jr. argues that by overemphasizing "climate mitigation", the UNFCCC and the COP agreements, such as the Kyoto Protocol (and more recently the Paris Agreement), have created a bias against investment in climate adaptation. The climate mitigation policies explicitly assume that climate change is primarily driven by greenhouse gas emissions, whereas climate adaptation policies often make sense regardless of the causes of climate change. With that in mind, it is worth noting that several recent studies have argued that the IPCC (2014) reports have underestimated the role of natural factors in recent climate change (and hence overestimated the role of human-caused factors) (Soon et al., 2015; Scafetta et al., 2019; Singh et al., 2020; Kumar et al., 2020).

11.2 What Is Renewable Energy?

Renewable energy is the energy obtained from regenerative resources and does not deplete over time. Renewable energy provides the world a chance to reduce carbon emissions, clean the air, and put humans on a more sustainable footing. It also offers countries around the world the chance to improve their energy security and stimulate economic development. Of the total world's energy consumption, renewable energy supplies approximately 18% (Kumar et al., 2010). Renewable energy accounted for about 19% of global final energy consumption in 2008, with conventional biomass accounting for 13% and hydroelectricity accounting for 3.2%. Renewable energy accounts for about 18% of global electricity production, with hydroelectricity accounting for 15% and emerging renewables accounting for 3%. Renewable energy can be used to substitute fossil fuels in four areas: electricity generation, hot water generation, transportation fuels, and rural (off-grid) energy services (Khalil, 2012).

11.3 The Significance of Renewable Energy Resources

Changing climate is a challenging task that the world faces today, and avoiding it would require significant improvements in the way we generate and use resources (IEA, 2002). Sources of energy are classified into three categories: fossil fuels such as natural gas, oil, coal (Valdez-Vazquez, 2010); renewable resources including geothermal, biomass, hydropower, solar, wind, marine energies; and nuclear resources (Demirbas, 2000; Gibran et al., 2014). Renewable energy sources met 14% of total global primary energy consumption at the turn of the century, and this figure had risen to nearly 17% by 2010, with a projected 50% increase by 2040 (Kalogirou, 2004).

Around three-quarters of the world's energy is generated by burning fossil fuels. But the burning of fossil fuels emits greenhouse gases (GHGs) that are now recognized as the gases responsible for changing climatic conditions (Lee, 2002). Renewable energy sources are called primary, low-risk, sustainable, and inexhaustible energy resources that can be effective because most renewable energy sources emit little or no greenhouse gases (Dincer, 2001; Bilgen, 2004; EIA, 2012). Greenhouse gas emissions connected with renewable energy, according to data, are very low (Field and Raupach, 2004).

11.4 How Does Renewable Energy Reduce GHG Emissions?

11.4.1 Solar Power

Solar thermal energy is the most promising renewable energy source, and it is distributed in both explicit and implicit forms. Table 11.1 shows solar energy can be used for a variety of thermal applications, including cooking, drying of the crops, water heating, solar thermal control, and solar photovoltaic systems (Panwar et al., 2011). Solar water heating systems are used for industrial and residential systems, whereas solar cookers can be used only on the domestic scale; therefore, for large-scale cooking, solar steam generating systems are employed. Besides cooking, this system can also clean clothes in the textile industry, hotels, etc. Apart from water heating and cooking, solar energy in India is also being used for air conditioning, refrigeration, water purification, and desalination. For textiles, chemicals, and plastics industries, solar power, and conventional boilers are potential power sources. These systems can also be used in conjunction with other conventional heating systems (Arora et al., 2010) (Figure 11.1).

The optimization of production processes towards low specific emissions and higher recycling rates will increase the benefit of the use of PV systems. Krauter and Ruther (2004) use thin-film technologies such as amorphous silicon, and others promise to reduce specific CO_2 emissions. In the case of Brazil also, off-grid applications and the substitution of diesel generating sets by PVs are examined: CO_2 reduction may reach 26,805 kg/kWp in that case. Doing these calculations, the compositions of the local grids and their CO_2 intensity at the time of PV grid injection have to be taken into account. During the operation time of a PV system, different kinds of power plants could be installed that might change the CO_2 intensity of the grid. In the future also, advanced technologies such as thin films have to be considered.

In line with the Paris Climate Change Accord for reducing greenhouse gases, South Korea has announced a strategy to expand its new 2030 energy industry (MOTIE, 2015). In the energy construction field, Zero Energy Building is to become mandatory for new buildings from 2025, with plans to start in the public sector (2020) and expand to the private sector (2025). Hence, it is essential to introduce

TABLE 11.1

Green House Gases Reduction by Different Technologies

Renewable Energy Resources	Methods used for GHG Reduction
Bioenergy Resource	• Ammonia • Pellet boiler (Domestic heat) • Chip boiler (District heat) • Small Electricity • Larger Electricity • Biochar • Biodiesel • Biomass gasifier • Gasifier based power generation system • Improved cook stoves
Geothermal	• Geothermal heat pumps • Enhanced geothermal systems (EGS) • Boiler and absorption chiller
Solar Thermal	• Glazed/unglazed solar collectors, • Flat plate solar collectors, • Evacuated tube solar collectors • Concentrating solar power systems (CSP) • Solar water heater
Renewable Natural Gas	• Anaerobic digestion • Thermal gasification
Biogas	• Flared biogas • CHP system
Hydropower	• Strategic dam planning

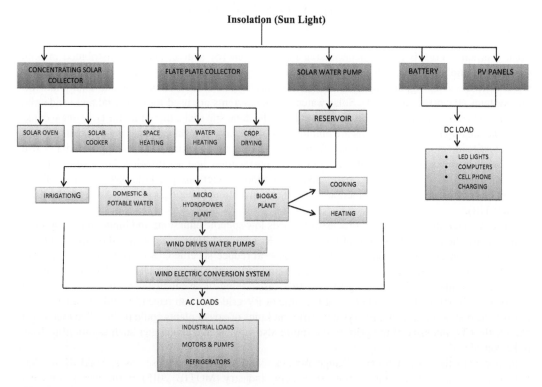

FIGURE 11.1 Role of renewable energy for reduction of greenhouse gases emission and achieving sustainable development.

renewable energy and high-factor insulation, high-performance windows, and heat recovery ventilation systems. South Korean apartments are mostly high-rise structures with more than ten floors. The only renewable energy devices that can be applied to high-rise buildings are limited to photovoltaic (PV) panels, solar collectors installed on the wall or balconies, or a photovoltaic thermal (PVT) hybrid panel combining both (Noh and Hwang, 2013; Kim et al., 2010).

Park et al. (2020) analyzed the effect of PV, Solar Thermal (ST), and PVT systems on greenhouse gas reduction using TRNSYS18. All three systems showed maximum CO_2 reductions at 35° facing south. PV, ST, and PVT showed CO_2 reductions of 67.4, 114.6, and 144.7 kg CO_2/m^2year, respectively. Compared to those values, when installed on a wall (slope of 90°), CO_2 reduction is about 35%–40% less and about 20% less at a slope of 75°. ST and PVT installed on the vertical wall have a greater greenhouse gas reduction effect than the PV installed at the optimal slope of 35°. Since the CO_2 reduction difference among SW, SE, and azimuthal S is within 10%, ST and PVT are recommended for installation on high-rise apartment structure walls or balconies with the azimuthal angle of ±45° with respect to south.

A CO_2 comprehensive balance within the life-cycle of a PV energy system requires careful examination of the CO_2 sinks and sources at the locations and under the conditions of production of each component, during transport, installation, and operation, as well as at the site of recycling. For the examples of Brazil and Germany, effective CO_2 reductions have been derived, also considering possible interchange scenarios for production and operation of the PV systems considering the carbon dioxide intensity of the local electricity grids. The substitution of diesel generating sets by PVs is examined in Brazil: CO_2 reduction may reach 26,805 kg/kWp in that case. Doing these calculations, the compositions of the local grids and their CO_2 intensity at the time of PV grid injection have to be taken into account. During the operation time of a PV system, different kinds of power plants could be installed that might change the CO_2 intensity of the grid. In the future, advanced technologies such as thin films have to be considered (Krauter and Ruther, 2004).

Large-scale GHG removal methods are when these GHGs can be mineralized by photocatalysis, using simple metal oxides like MgO or ZnO, cheap TiO_2 derivatives, and zeolites. Solar Chimney Power Plants (SCPPs) with Photocatalytic Reactors (PCRs)supported by the infrastructure of a SCPP is proposed as advisable to transform CH_4 into CO_2, which has GWP>10 times and transform several CFCs, HFCs into CO_2, $CHCl$, and HF. It has a capacity to neutralize the haloacids by-products which reduce the atmospheric concentrations of CH_4, N_2O, and halocarbons, thus achieving GHG removal; SCPP PCR is an emissions technology that might help to fight global warming and tackle climate change by preventing or removing nearly 1,600 kton seq-CO_2yr^{-1}(for each 200MW plant). Combined TiO_2-photocatalysis with SCPPs helps to cleanse the atmosphere of non-CO_2 GHGs. Worldwide installation of 50,000 SCPPs, each of capacity 200MW, would generate accumulative 34PWh of renewable electricity by 2050, taking into account construction time (Mills et al., 2016; Engel et al., 2015).

11.4.2 Hydropower

There is a clear connection between hydropower and climate change. On the one side, hydropower is a valuable renewable energy source that helps to reduce greenhouse gas emissions and mitigate global warming. Climate change will alter river discharge, resulting in changes in water availability, water regularity, and hydropower generation (Berga, 2016). As compared to coal-fired power plants, hydropower eliminates the emission of approximately 3 GT CO_2 annually or around 9% of global annual CO_2 emissions. Hydropower, in particular, is an electricity source that emits few greenhouse gases. The World Energy Council (WEC) estimates that CO_2 emissions per GWh for run-of-the-river hydropower are 3–4t, and 10–33t for hydropower with a reservoir; these figures are roughly 100 times lower than emissions from conventional thermal power (World Energy Council, 2004). Turbines are constructed for an optional flow of water (Forsund, 2015).

Hydropower emits almost no particulate matter, can be upgraded easily, and can store electricity for several hours (Hamann, 2016). Its development has a very important role in future mitigation scenarios of climatic change. In the International Energy Agency (IEA) scenario of an emissions peak at 450ppm with a maximum increment in temperature of 2°C, which is the generally accepted scenario, the capacity of hydropower generation should increase 70% by 2030 and 100% by 2050 (International Energy Agency, 2012).

Hydropower has several advantages over most other renewable resources, such as high degree of reliability, high efficiency, proven technology, and low operating and maintenance costs, and it does not generate waste products that contribute to acid rain and greenhouse gas emissions. Most hydropower plants are typically placed near reservoirs that provide water, flood control, and recreational opportunities for the region (Liu, 2013). Climate change is expected to have a global impact of less than 0.1% on the current hydropower production system. Hydropower production does not emit greenhouse gases, so it is often referred to as a renewable energy source. The construction of dams, dikes, and weirs affects the river body's ecology, primarily by altering its hydrologic features and interfering with the ecological consistency of sediment transport and fish movement (Sathayeet al., 2011). More recently, the International Renewable Energy Agency's (IRENA) global Renewable Energy Roadmap (REMAP 2030), in line with the UN Secretary-General "Sustainable Energy for All" (SE4ALL) scenario, which aims to double the global share of renewable energy as described in "Doubling the Global Share of Renewable Energy: A Road Map to 2030", necessitated that it requires around 2,200 GW of total hydropower capacity to achieve its targets. This assumes an additional 500 GW of hydropower capacity to be built, in addition to the IEA projections (IRENA, 2013).

11.4.3 Wind Energy

The production of electricity from large turbines located onshore (on land) or offshore (in the sea or freshwater) is the primary application for climate change mitigation (Asumadu-Sarkodie & Owusu, 2016). Onshore wind power systems have been developed and installed on a wide scale right now (Edenhofer et al., 2011). Wind turbines convert the energy of wind into electricity. Wind energy decreases greenhouse gas emissions and lowers power costs by using turbines that generate energy and electricity when driven by the wind. All that is required for the turbines to operate is wind, that is, simply air moving in a natural direction, and as we know, air is found everywhere. Wind power is a natural solution that encourages social change by denying the pessimistic prediction of a planet depleted of oil and fuels (Bull, 2001). Lenzen and Munksgaard (2002) identified that, over the life cycle of a wind turbine, the amount of energy used and the amount of CO_2 emitted is reduced. They discovered that considering the price of energy needed for the components, small wind turbines, those generating 1 kW, need a significantly higher amount of energy over their life cycle than larger ones. While wind turbines may not be practical in all places, they may be preferable to natural gas or coal-fired generators in areas where the wind is more favourable (Liberman, 2003).

Increasing the wind energy installation capacity plays a crucial role in climate change mitigation. However, wind energy is also susceptible to global climate change. Some changes associated with climate evolution will most likely benefit the wind energy industry. Wind energy for electricity generation is now an established, efficient, and pollution-free technique that is commonly used in certain parts of the world (Balat, 2009). The electricity produced by the wind energy industry could save several billion barrels of oil and avoid many million tons of carbon and other emissions (Thomas and Urquhart, 1996). As shown in Table 11.2, at a mean wind speed of 4.5 m/s, the estimated value of net annual CO_2 emission mitigation potential is the lowest (2,874 kg) for the GM-II model and the highest (7,401 kg) for the SICO model in the case of diesel substitution. Similarly, for the case of electricity substitution for the same wind speeds, it is estimated at 2,194 and 5,713 kg, respectively, for the two models mentioned above (Kumar and Kandpal, 2007a).

11.4.4 Geothermal Energy

Geothermal energy, one of the renewable energies, has great importance for some parts of the world. In comparison to wind and solar energy, geothermal energy has the benefit of being available 24 hours a day, 365 days a year. Ngo and Natowitz (2009) reported on data collected from 73% of the geothermal power plants. The CO_2 emissions emitted by geothermal resources are expected to be 55 g/kWh. Such value can be reduced to zero if geothermal fluid is re-injected into the earth. In 2011, a maximum of

TABLE 11.2

Renewable Energy Sector and Its Potential to Reduce CO_2

Energy Sources	CO_2 Emission Reduction Capacity	References
Solar energy		
(i) Solar cooking	• 38.4 million tons of carbon dioxide per year	Nandwani (1996)
(ii) Solar water Heater	• 1,237 kg of CO_2 emissions in a year	Kumar and Kandpal (2007a) Jyotirmay et al. (2002)
(iii) Solar-drying technology	• 463 kg of carbon dioxide in life cycle embodied	Kumar and Kandpal (2005)
(iv) Solar photovoltaic system	• 1.8 kWp solar photovoltaic pump at an average solar radiation of 5.5 $kWhm^{-2}$ is about 2,085 kg from diesel operated pumps • about 1,860 kg from petrol operated pumps.	Kumar and Kandpal (2007b)
Wind energy	• CO_2 emission mitigation potential is the lowest (2,874 kg) • for GM-II model • Highest (7,401 kg) for SICO model in the case of diesel substitution.	Kumar and Kandpal (2007a)
Bioenergy		
(i) Biogas	• 9.7 tons CO_2 equiv. $year^{-1}$	Pathak et al. (2009)
(ii) Biodiesel	• 79,782 tons of CO_2 per year	Panwar et al. (2009a)
(iii) Improved cookstoves	• Single stove can save about 700 kg of fuel wood per year and at the same time it reduces the CO_2 emission by 161 kg per year	Panwar et al. (2009a)

24 countries used geothermal power plants, with a total installed capacity of 11 GW. Geothermal energy, or heat originating from the earth's interior, has three applications: geothermal heat pumps, electricity generation, and heating systems (and direct use). Since geothermal energy plants do not use power, they are both environmentally friendly and renewable.

In order to measure the impact of geological conditions on the climate, Frick et al. (2010) used a life cycle analysis tool to look at geothermal power production from an advanced geothermal system (EGS) with low-temperature reservoirs. The findings indicate that its materials and energy inputs determine a geothermal binary power plant's life cycle. As a result, gaining access to a reservoir with minimal drilling is a critical component of minimizing environmental consequences. Also, less favourable geothermal heat and power generation have been shown to lead to a sustainable energy system. Saner et al. (2010) examined energy use and greenhouse gas emissions. They also used a life cycle assessment to look at the environmental effects of a ground source heat pump (GSHP) that was installed to extract geothermal energy. As compared to traditional heating systems, the findings show a 31%–88% reduction in emissions.

Carbon dioxide emissions from high-temperature geothermal fields used for electricity generation in the world range from 13 to 380 g/a kilowatt-hour, which is lower than that from fossil fuel power plants. Sulphur emissions from geothermal power plants are also much lower than those from fossil fuel power plants. In Reykjavik, Iceland, the carbon dioxide level in low-temperature water is lower than that of cold groundwater. With its existing technology and plentiful resources, geothermal energy has the potential to make a major contribution to reducing greenhouse gas emissions. However, policymakers must adopt policies and strategies to make geothermal energy systems more efficient than traditional energy systems (U.S. DOE (Department of Energy), OGT, 1998).

Geothermal technologies produce little or no greenhouse gas emissions since no burning processes are involved. Geothermal development estimates for 2050 indicate that power generation could mitigate CO_2 emissions by 100s of Mt/yr, direct use >300 Mt/yr, most of it by geothermal heat pumps. These need electricity; its source must be considered. When complemented by measures in improved construction solutions like efficient thermal isolation to reduce the energy consumption of buildings, the geothermal heat pump systems to provide space heating and cooling and domestic hot water can and will contribute in the future significantly to avoiding or reducing CO_2 emissions. The environmental benefits of geothermal development are obvious: increasing development can help to mitigate the effects of global warming. The societal benefits are developing in parallel: the positive effects are also increasing, regionally and globally (Rybach, 2010).

11.4.5 Biomass

Today, the use of biomass as a source of energy has increased, accounting for around 14% of global total energy consumption. As per the estimates, by the end of the year 2050, biomass could account for 15%–50% of the world's primary energy use. In comparison to fossil fuels, biomass energy is a type of renewable energy that, in theory, may not contribute carbon dioxide, a potent greenhouse gas, to the atmosphere (Ramachandra et al., 2004; Senneca, 2006). Most bioenergy systems will help mitigate climate change when used instead of fossil fuels, and bioenergy emissions are reduced dramatically. GHG savings can be hampered by high nitrous oxide emissions from feedstock production and the use of fossil fuels (especially coal) in the biomass conversion process. Best fertilizer management practices, process integration to reduce losses, surplus heat use, and other low-carbon energy sources as process fuel are all options for lowering GHG emissions. When additional biomass feedstock is used for process energy in the conversion process, nevertheless, the displacement efficiency (GHG emissions relative to carbon in biomass) can be low unless the displaced energy is produced from coal. The displacement efficiency will be high if the biomass feedstock can generate both liquid fuel and electricity.

Biomass energy has the potential to significantly minimize greenhouse gas emissions, reliance on foreign oil, landfills, and eventually, help local agricultural and forest-product industries. Papermill residue, lumber mill scrap, and municipal waste are the most popular biomass power feedstocks. Nowadays, corn grain for ethanol and soybeans for biodiesel are the two most popular feedstocks for biomass fuels. Long-term strategies for biomass energy production include increasing and utilizing dedicated energy crops like fast-growing trees and grasses, as well as algae. These feedstocks can be grown on land that is not suitable for intensive food crops. Another advantage of biomass is that, like crude oil, it can be converted into a variety of useful fuels, chemicals, materials, and goods (http://www.nrel.gov/learning/re_biomass.html). Biogas technology offers a great way to reduce greenhouse gas emissions and global warming by replacing firewood for heating, kerosene for lighting and cooking, and chemical fertilisers with biogas. Under the clean development mechanism, the global warming mitigation potential of a family-size biogas plant was 9.7 tons CO_2 Equiv. year^{-1} and with the current price of US \$10 tons^{-1} CO_2 Equiv., carbon credit of US \$97 year^{-1} could be earned from such reduction in greenhouse gas emission (Pathak et al., 2009).

The transport industry is the largest producer of greenhouse gases, and biodiesel has the potential to reduce the emissions of greenhouse gases from vehicles. The use of biodiesel also reduces the particulate matter released into the atmosphere as a result of burning fuels, providing potential benefits to human health (Beer et al., 2007). A study was reported in the Indian context that if 10% of the total production of the castor seed oil is transesterified into biodiesel, then about 79,782 tons of CO_2 emission can be saved on an annual basis. The CO_2 released during the combustion of biodiesel can be recycled through the next crop production. Therefore, there is no additional burden on the environment (Panwar et al., 2010). The potential of three fuel types: ethanol, methanol, and vegetable oil have to release less carbon monoxide and sulphur dioxide than gasoline and diesel fuels; however, because the production of most of these biofuels requires more total fossil energy than they produce as a biofuel, they contribute to air pollution and global warming (Pimentel et al., 2002).

11.4.6 Improved Cookstoves

Smith et al. (2000) studied the combustion process in the traditional cooking stove and identified it as non-ideal and favouring incomplete combustion. The fine and ultrafine particles released from incomplete combustion (PIC) and inefficient combustion by traditional cookstoves have more global warming potential (GWP) than CO_2. The improved cookstoves provide a better kitchen environment to rural women and improve their health standards. At the same time, it also reduces the fuel collection burden for them. Improvements in households biomass burning stoves potentially bring three kinds of benefits: (1) reduced fuel demand, with economic and time-saving benefits to the household and increased sustainability of the natural resources base; (2) reduced human exposure to health-damaging air pollutants; and (3) reduced emission of the greenhouse gases that are thought to increase the probability of global climate change. Single stoves have the capacity to save about 700 kg of fuelwood per year by reducing the CO_2 emission by 161 kg per year (Smith, 1994).

11.5 GHG Reduction Costs

Accounting for greenhouse gas emissions from all phases of the project (construction, operation, and decommissioning) is called a life cycle approach. Normalizing the life cycle emissions with electrical generation allows for a fair comparison of the different generation methods per gigawatt-hour basis. The lower the value, the less GHG emissions are emitted. A comprehensive response to that problem would include a collection of strategies: research to understand the scientific processes at work better and to develop technologies to address them; measures to help the economy and society adapt to the projected warming and other expected changes; and efforts to reduce emissions, averting at least some of the potential damage to the environment and attendant economic losses. Those strategies would all present technological challenges and entail economic costs. Reducing emissions would impose a burden on the economy because it would require lessening fossil fuels and altering land use patterns. Wood waste (with methane reduction) and landfill gas have the lowest cost per unit of GHG reduction from renewable energy sources. Wind and solar energy systems do not emit CO_2, but they have relatively high capital costs and low power factors, making them less desirable for reducing GHG emissions at the moment (Tennessee Valley Authority, 2003).

The costs and complexities of incorporating increasing shares of renewable energy into an established energy supply system are determined by system components such as existing renewable energy resources, the current share of renewable energy, and how the system evolves and expands in the future. Renewable energy incorporation is contextual, site-specific, and dynamic, either for power, heating, cooling, gaseous fuels, or liquid fuels. Due to the extreme geographical distribution and defined remote locations of many renewable energy resources, a combination of inexpensive and efficient communications systems and technologies and smart metres will be needed. Maintaining system reliability becomes more difficult and expensive as the percentage of partially dispatchable renewable energy electricity grows (Moomaw, 1991; Kempton and Tomic, 2005).

11.6 Conclusion

Energy is a must in our daily lives in order to advance human progress, which leads to increased economic growth and productivity. Renewable energy technologies are regarded as environmentally friendly energy sources, and the most efficient use of these resources has the least amount of environmental effects, generates the least amount of secondary waste, and is long-term dependent on long-term economic and social societal needs. Increased renewable energy availability, on the other hand, will enable us to substitute carbon-intensive sources of energy and reduce global warming emissions. Solar drying of agricultural produce has a lot of potential for saving energy in developing countries. Biodiesel from nonedible vegetable oil reduces carbon dioxide emissions and petroleum consumption when used in place of conventional diesel (Carraretto et al., 2004). Biodiesel is technically competitive with or offers technical advantages compared to conventional petroleum diesel fuel. The fact that we have been utilizing fossil fuels is terrifying, and what is even scarier is that consumption has increased in recent decades. Pollution in the atmosphere has disastrous effects, including global warming; thus, we must protect the world by integrating clean, environmentally sustainable energy sources into our daily lives. This chapter emphasizes the importance of renewable energy in reducing greenhouse gas emissions, combating climate change, and replacing fossil fuels.

REFERENCES

Arora, D. S., S. Busche, S. Cowlin, T. Engelmeier, J. Jaritz, A. Milbrandt and S. Wang. 2010. Indian renewable energy status report: Background report for DIREC 2010 (No. NREL/TP-6A20–48948). National Renewable Energy Lab.(NREL), Golden, CO (United States).

Asumadu-Sarkodie, S. and P. A. Owusu. 2016. The potential and economic viability of wind farms in Ghana. *Energy Sources, Part A: Recovery, Utilization, and Environmental Effects* 38: 695–701.

Balat, M. 2009. A review of modern wind turbine technology. *Energy Sources, Part A* 31: 1561–1572.

Banos, R., F. Manzano-Agugliaro, F. G. Montoya, C. Gil, A. Alcayde and J. Gomez. 2011. Optimization methods applied to renewable and sustainable energy: A review. *Renewable and Sustainable Energy Reviews* 15: 1753–1766.

Beer, T., T. Grant and P. K. Campbell. 2007. The greenhouse and air quality emissions of biodiesel blends in Australia. CSIRO Marine and Atmospheric Research. http://www.csiro.au/resources/ pf13o.html (Accessed November 20, 2009).

Berga, L. 2016. The role of hydropower in climate change mitigation and adaptation: A review. *Engineering* 2: 313–318.

Bilgen, S., K. Kaygusuz and A. Sari. 2004. Renewable energy for a clean and sustainable future. *Energy Sources* 26: 1119–1129.

Bull, S. R. 2001. Renewable energy today and tomorrow. *Proceedings of the IEEE* 89: 1216–1226.

Cao, L. and K. Caldeira. 2010. Atmospheric carbon dioxide removal: Long-term consequences and commitment. *Environmental Research Letters* 5: 024011.

Carraretto, C., A. Macor, A. Mirandola, A. Stoppato and S. Tonon. 2004. Biodiesel as alternative fuel: Experimental analysis and energetic evaluations. *Energy* 29: 2195–2211.

Connolly, R., M. Connolly, R. M. Carter and W. Soon. 2020. How much human-caused global warming should we expect with business-as-usual (BAU) climate policies? A semi-empirical assessment. *Energies* 13: 1365.

Demirbas, A. 2000. Recent advances in biomass conversion technologies. *Energy Education Science and Technology* 6: 19–40.

Dincer, I. 2001. Environmental issues: I-energy utilization. *Energy Sources* 23: 69–81.

Edenhofer, O., R. Pichs-Madruga, Y. Sokona, K. Seyboth, P. Matschoss, S. Kadner, T. Zwickel, P. Eickemeier, G. Hansen, S. Schlömer and C. von Stechow. 2011. IPCC special report on renewable energy sources and climate change mitigation. *Prepared By Working Group III of the Intergovernmental Panel on Climate Change*. Cambridge, UK: Cambridge University Press.

Energy Information Agency (EIA) (2012). How much of the U.S. carbon dioxide emissions are associated with electricity generation. https://www.nrc.gov/docs/ML1408/ML14086A551.pdf.

Engel, A., A. Glyk, A. Hulsewig, J. Grobe, R. Dillert and D. W. Bahnemann. 2015. Determination of the photocatalytic deposition velocity. *Chemical Engineering Journal* 261: 88–94.

Fay, J. A. and D. S. Golomb. 2002. *Energy and the Environment*. New York: Oxford University Press. http://www.oup-usa.org.

Field, C. B. and M. R. Raupach. eds. 2004. *The Global Carbon Cycle: Integrating Humans, Climate, and the Natural World, Scientific Committee on Problems of the Envrionment (SCOPE) 62: 526.* Washington, D.C.: Island Press.

Forsund, F. R. 2015. *Hydropower Economics*. Boston, MA: Springer.

Frick, S., M. Kaltschmitt and G. Schroder. 2010. Life cycle assessment of geothermal binary power plants using enhanced low-temperature reservoirs. *Energy* 35: 2281–2294.

Friedlingstein, P., M. W. Jones, M. O'sullivan, R. M. Andrew, J. Hauck, G. P. Peters and S. Zaehle. 2019. Global carbon budget 2019. *Earth System Science Data* 11: 1783–1838.

Gibran, S. A., H. C. Victor, L. C. Diana, R. Diaz-Chavez, N. Scarlat, J. Mahlknecht and R. Parra. 2014. Renewable energy research progress in Mexico: A review. *Renewable and Sustainable Energy Reviews* 32: 140–153.

Hamann, A. and H. Gabriela. 2016. Integrating variable wind power using a hydropower cascade. *Energy Procedia* 87: 108–115. 5th International Workshop on Hydro Scheduling in Competitive Electricity Markets

Hotchkiss, E. R., R. O. Hall Jr., R. A. Sponseller, D. Butman, J. Klaminder, H. Laudon and J. Karlsson. 2015. Sources of and processes controlling CO_2 emissions change with the size of streams and rivers. *Nature Geoscience* 8: 696–699.

IEA. 2002. Beyond Kyoto—Energy Dynamics and Climate Stabilistation. International Energy Agency, OECD, Paris. http://philibert.cedric.free.fr/Downloads/Beyond%20Kyoto_NS.pdf.

International Agency Energy. Technology roadmap: hydropower. Paris: International Energy Agency; 2012. https://webstore.iea.org/technology-roadmap-hydropower.

International Renewable Energy Agency (IRENA). 2013. IRENA REMAP 2030: Doubling the global share of renewable energy, a roadmap to 2030. Abu Dhabi: IRENA.https://irena.org/-/media/Files/IRENA/Agency/Publication/2013/IRENA-REMAP-2030-working-paper.pdf.

International Renewable Energy Agency (IRENA). IRENA REMAP 2030: doubling the global share of renewable energy, a roadmap to 2030. Abu Dhabi: IRENA; 2013. https://irena.org/publications/2013/Jan/Doubling-the-Global-Share-of-Renewable-Energy-A-Roadmap-to–2030.

IPCC. 2013. Climate change 2013: The physical science basis. In *Contribution of Working Group I Contribution to the Fifth Assessment Report of the Intergovernmental Panel on Climate Change. Intergovernmental Panel on Climate Change (IPCC)*. Eds. T. F. Stocker, D. Qin, G. K. Plattner, M. Tignor, S. K. J. Allen, A. Boschung, Y. Nauels, V. Xia, Bex, and P. M. Midgley. Cambridge, UK and New York, USA: Cambridge University Press.

Jyotirmay, M., N. K. Kumar and H. J. Wagner. 2002. Energy and environmental correlation for renewable energy systems in India. *Energy Sources, Part A: Recovery, Utilization, and Environmental Effects* 24:19–26.

Kalogirou, S. A. 2004. Solar thermal collectors and applications. *Progress in Energy and Combustion Science* 30: 231–295.

Kempton, W. and J. Tomic. 2005. Vehicle-to-grid power implementation: From stabilizing the grid to supporting large-scale renewable energy. *Journal of Power Sources* 144: 280–294.

Khalil, E. E. 2012. The role of solar and other renewable energy sources on the strategic energy planning: AFRICA's status & views. *Ashrae Transactions* 118: 64–72.

Kim, J. S., E. J. Lee and J. H. Hwang. 2010. A study on electric capacity and CO_2 by the roof top PV system of the industrial building in Korea. *Journal of the Korean Solar Energy Society* 30: 131–136.

Krauter, S. and R. Ruther. 2004. Considerations for the calculation of greenhouse gas reduction by photovoltaic solar energy. *Renewable Energy* 29: 345–355.

Kumar, A. and T. C. Kandpa. 2005. Solar drying and CO_2 emissions mitigation: Potential for selected cash crops in India. *Solar Energy* 78:321–9.

Kumar, A. and T. C. Kandpal. 2007a. CO_2 emissions mitigation potential of some renewable energy technologies in India. *Energy Sources, Part A* 29: 1203–1214.

Kumar, A. and T. C. Kandpal. 2007b. Potential and cost of CO_2 emissions mitigation by using solar photovoltaic pumps in India. *International Journal of Sustainable Energy* 26:159–66.

Kumar, A., K. Kumar, N. Kaushik, S. Sharma and S. Mishra. 2010. Renewable energy in India: current status and future potentials. *Renewable and Sustainable Energy Reviews* 14: 2434–2442.

Kumar, A., P. Kumar, H. Singh and N. Kumar. 2020. Adaptation and mitigation potential of roadside trees with bio-extraction of heavy metals under vehicular emissions and their impact on physiological traits during seasonal regimes, Urban Forestry & Urban Greening https://doi.org/10.1016/j.ufug.2020.126900.

Lee, J. S., D. I. Won, W. J. Jung, H. J. Son, C. Pac and S. O. Kang. 2017. Widely controllable syngas production by a dye-sensitized TiO_2 hybrid system with ReI and CoIII catalysts under visible-light irradiation. *AngewandteChemie International Edition* 56: 976–980.

Lee, R. 2002. Environmental impacts of energy use. Energy: science, policy, and the pursuit of sustainability. ed. R. Bent, L. Orr, R, Baker, 77. Washington: Island Press.

Lenzen, M. and J. Munksgaard. 2002. Energy and CO_2 life-cycle analyses of wind turbines—review and applications. *Renewable Energy* 26: 339–362.

Liberman, E. J. 2003. A life cycle assessment and economic analysis of wind turbines using monte carlo simulation, Air Force Institute of Technology, Wright-Patterson Air Force Base, 2003. https://apps.dtic.mil/dtic/tr/fulltext/u2/a415268.pdf.

Liu, G., E. D. Larson, R. H. Williams, T. G. Kreutz and X. Guo.2011. Making Fischer-Tropsch fuels and electricity from coal and biomass: Performance and cost analysis. *Energy & Fuels* 25: 415–437.

Liu, Y. H., F. K. Qian and W. T. Wang. 2013. Research of adaptive technology framework of addressing climate change. *Chinese Journal of Population Resources and Environment* 23:1–6.

Matthews, H. D. and K. Caldeira. 2008. Stabilizingclimaterequiresnear-zeroemissions. *Geophysical Research Letters* 35: L04705. doi:10.1029/2007GL032388.

Menyah, K. and Y. Wolde-Rufael. 2010. CO_2 emissions, nuclear energy, renewable energy and economic growth in the US. *Energy Policy* 38: 2911–2915.

Mills, A., L. Burns, C. O'Rourke and S. Elouali. 2016. Kinetics of the photocatalysed oxidation of NO in the ISO 22197 reactor. *Journal of Photochemistry and Photobiology A: Chemistry* 321:137–142.

Moomaw, W. R. 1991. Photovoltaics and materials science: Helping to meet the environmental imperatives of clean air and climate change. *Journal of Crystal Growth* 109: 1–11.

MOTIE (Ministry of Trade, Industry, and Energy). 2015. To Respond to the New Climate System 2030 New Industry of Energy Expansion Strategy; Ministry of Trade, Industry and Energy: Seoul, Korea.

Nandwani, S. S. 1996. Solar cookers cheap technology with high ecological benefits. *Ecological Economics* 17:73–81.

Ngo, C. and J. Natowitz. 2009. *Our Energy Future: Resources, Alternatives and the Environment.* Volume 11 of Wiley Survival Guides in Engineering and Science. Hoboken, NJ: Wiley.

Noh, J. Y. and D. K. Hwang. 2013. Solar hot water system and photovoltaic system performance analysis in an apartment. In *Proceedings of the SAREK Summer Annual Conference*, Yongpyong, Korea, (June): 339–342.

O'Neill, B. C., M. Oppenheimer, R. Warren, S. Hallegatte, R. E. Kopp, H. O. Portner and G. Yohe. 2017. IPCC reasons for concern regarding climate change risks. *Nature Climate Change* 7: 28–37.

Panwar, N. L., H. Y. Shrirame, N. S. Rathore, S. Jindal and A. K. Kurchania. 2010. Performance evaluation of a diesel engine fueled with methyl ester of castor seed oil. *Applied Thermal Engineering* 30: 245–249.

Panwar, N. L., A. K. Kurchania and N. S. Rathore. 2009a. Mitigation of greenhouse gases by adoption of improved. *Mitigation and Adaptation Strategies for Global Change* 14:569–78.

Panwara, N. L., S. C. Kaushik and S. Kothari. 2011. Role of renewable energy sources in environmental protection: A review. *Renewable and Sustainable Energy Reviews* 15: 1513–1524.

Park, C. H., Y. J. Ko, J. H. Kim and H. Hong. 2020. Greenhouse gas reduction effect of solar energy systems applicable to high-rise apartment housing structures in South Korea. *Energies* 13: 2568.

Pathak, H., N. Jain, A. Bhatia, S. Mohanty and N. Gupta. 2009. Global warming mitigation potential of biogas plants in India. *Environmental Monitoring and Assessment* 157: 407–418.

Pielke Jr., R. A. 2005. Misdefining "climate change": Consequences for science and action. *Environmental Science & Policy* 8: 548–561.

Pielke, R., G. Prins, S. Rayner and D. Sarewitz. 2007. Lifting the taboo on adaptation. *Nature* 445: 597–598.

Pimentel, D., M. Herz, M. Glickstein, M. Zimmerman, R. Allen, K. Becker and T. Seidel. 2002. Renewable Energy: Current and Potential Issues Renewable energy technologies could, if developed and implemented, provide nearly 50% of US energy needs; this would require about 17% of US land resources. *Bioscience* 52: 1111–1120.

Ramachandra, T. V., G. Kamakshi and B. V. Shruthi. 2004. Bioresource status in Karnataka. *Renewable and Sustainable Energy Reviews* 8:1–47.

Reddy, B. S. and G. B. Assenza. 2009. The great climate debate. *Energy Policy* 37: 2997–3008.

Rybach, L. 2010. CO_2 emission mitigation by geothermal development especially with geothermal heat pumps. *Proceedings of the World Geothermal Congress* 2010, Bali, (April): 1–4.

Sakata, M., H. G. Phan and S. Mitsunobu. 2017. Variations in atmospheric concentrations and isotopic compositions of gaseous and particulate boron in Shizuoka City, Japan. *Atmospheric Environment* 148: 376–381.

Saner, D., R. Juraske, M. Kubert, P. Blum, S. Hellweg and P. Bayer. 2010. Is it only CO_2 that matters? A life cycle perspective on shallow geothermal systems. *Renewable and Sustainable Energy Reviews* 14: 1798–1813.

Sathaye, J., O. Lucon, A. Rahman, et al. 2011. Renewable energy in the context of sustainable development. In *Renewable Energy Sources and Climate Change Mitigation. IPCC Special Report on Renewable Energy Sources and Climate Change Mitigation*, eds. O. Edenhofer, R. Pichs-Madruga, Y. Sokona, K. Seyboth, P. Matschoss, S. Kadner, T. Zwickel, P. Eickemeier, G. Hansen, S. Schlomer, C. Von Stechow, 707–789. Cambridge, United Kingdom and New York, NY, USA: Cambridge University Press.

Scafetta, N., R. C. Willson, J. N. Lee and D. L. Wu. 2019. Modeling quiet solar luminosity variability from TSI satellite measurements and proxy models during 1980–2018. *Remote Sensing* 11: 2569.

Senneca, O. 2006. Kinetics of pyrolysis, combustion and gasification of three biomass fuels. *Fuel Process Technology* 88: 87–97.

Singh, H., M. Yadav, N. Kumar, A. Kumar and M. Kumar 2020. Assessing adaptation and mitigation potential of roadside trees under the influence of vehicular emissions: A case study of Grevillea robusta and Mangifera indica planted in an urban city of India. *PLoS ONE* 15(1): e0227380.

Smith, K. R. 1994. Health, energy, and greenhouse-gas impacts of biomass combustion in household stoves. *Energy for Sustainable Development* 1: 23–29.

Smith, K. R., R. Uma, V. V. N. Kishore, J. Zhang, V. Joshi and M. A. K. Khalil. 2000. Greenhouse implications of household stoves: An analysis for India. *Annual Review of Energy and the Environment* 25: 741–763.

Soon, W., R. Connolly and M. Connolly. 2015. Re-evaluating the role of solar variability on Northern Hemisphere temperature trends since the 19th century. *Earth-Science Reviews* 150: 409–452.

Stocker, T. F., D. Qin, G.K. Plattner, L.V. Alexander, S. K. Allen, N. L. Bindoff, F. M. Bréon, J. A. Church, U. Cubasch, S. Emori and P. Forster. 2013. Technical summary. In *Climate change 2013: the physical science basis. Contribution of Working Group I to the Fifth Assessment Report of the Intergovernmental Panel on Climate Change*, 33–115. Cambridge University Press.

Tennessee Valley Authority. 2003. The Role of Renewable Energy in Reducing Greenhouse Gas Buildup. https://www.nrc.gov/docs/ML1217/ML12170A464.pdf.

Thomas, B. G. and J. Urquhart. 1996. Wind energy for the 1990s and beyond. *Journal of Energy Conversion and Management* 37:1741–52.

Tian, H., C. Lu, P. Ciais, A. M. Michalak, J. G. Canadell, E. Saikawa, D. N. Huntzinger, K.R. Gurney, S. Sitch, B. Zhang and J. Yang. 2016. The terrestrial biosphere as a net source of greenhouse gases to the atmosphere. *Nature* 531(7593): 225–228.

U.S. DOE OGT. 1998. Strategic Plan for the Geothermal Energy Program. U.S. Department of Energy, Office of Geothermal Technologies, Washington, pp. 23.

UNFCCC. 2015. Paris Agreement, United Nations Framework Convention on Climate Change. FCCC/CP/2015/L.9. United Nations.NewYork; http://unfccc.int/. (Accessed December 12, 2015).

Valdez-Vazquez, I., J. A. Acevedo-Benitez and C. Hernandez-Santiago. 2010. Distribution and potential of bioenergy resources from agricultural activities in Mexico. *Renewable and Sustainable Energy Reviews* 14: 2147–2153.

Vichi, M., A. Navarra and P. G. Fogli. 2013. Adjustment of the natural ocean carbon cycle to negative emission rates. *Climatic Change* 118:105–118.

Wang, Q. and R. Li. 2016. Journey to burning half of global coal: Trajectory and drivers of China's coal use. *Renewable and Sustainable Energy Reviews* 58: 341–346.

Wen, X. Y., G. L. Tang, S. W. Wang et al. 2011.Comparison of global mean temperature series. *Advance in Climate Change Research* 2:187–192.

Williams, J. H., A. DeBenedictis, R. Ghanadan, A. Mahone, J. Moore, W. R. Morrow and M. S. Torn. 2012. The technology path to deep greenhouse gas emissions cuts by 2050: The pivotal role of electricity. *Science* 335: 53–59.

World Energy Council. 2004. Comparison of energy systems using life cycle assessment: a special report of the World Energy Council. London: World Energy Council. https://www.worldenergy.org/assets/downloads/PUB_Comparison_of_Energy_Systens_using_lifecycle_2004_WEC.pdf.

12

Wind Energy Technologies: Current Status and Future Prospective

Pankaj Singh Rawat, R.C. Srivastava, and Gagan Dixit

CONTENTS

12.1 Introduction

The use of renewable energy dates back to the advent of humankind. Centuries ago, our ancestors made one of the three most significant discoveries that transformed the destiny of entire planet. It was the discovery of fire. This influenced the future of human race and assisted the other two discoveries—wheel and agriculture—to shape the world that we see today. Initially, humans used renewable energy sources like hydropower for the purpose of irrigation and to operate various mechanical devices and wind energy for sailboats and ships. This dawned the era of transcontinental trade and commerce. However, the situation did not last long, and by the second half of the 18th century, fossil fuels took over all other energy resources. Fossil fuels are the carbon-rich anaerobically decomposed remains of plants, animals, and other organisms that are buried beneath the layers of rocks and sediments for centuries and release energy on combustion. These are coal, oil, and natural gases. The commercial mining of coal in the world began in the 17th century. Very soon, coal became the preferred fuel of choice for steam engines since it was cheaper and had higher calorific value than its traditional counterparts like firewood and charcoal.

DOI: 10.1201/9781003175926-12

Various reports suggest that petroleum, coal, and natural gases have been the world's primary energy source, accounting for up to 85% of the share. Renewable energy sources, on the other hand, have been accounting for a mere ~5%. Excessive exploitation of fossil fuels has triggered a globally threatening phenomenon of climate change resulting from global warming caused by greenhouse gases. When fossil fuels burn, they emit greenhouse gases (viz. carbon dioxide, nitrous oxide, methane, ozone, chlorofluorocarbons, hydro-fluorocarbons, etc.). These greenhouse gases trap the heat in our atmosphere, thereby raising the earth's temperature globally and resulting in climate change (Kumar et al., 2019, 2020a). With the peril of global warming and climate change, the tide has strongly turned in favour of the utilization of renewable energy as an alternate energy source. With increasing environmental awareness, governments and legislative authorities are considering a more sustainable energy paradigm of utilizing renewable energy resources like wind, hydro, and solar energies (Stokes and Warshaw, 2017; Bulut and Muratoglu, 2018; Bersalli et al., 2020; Wang et al., 2020c; Majid, 2020; Khan et al., 2020; Li et al., 2020a).

Wind energy and other renewable forms of energy are one of the primordial energy forms known to man. It has played a significant role in shaping the history of human civilization. Traditionally, wind energy was utilized for sailing ships, leading to the discovery of various trade routes and the first stage of globalization. Harnessing and utilizing wind power as a resource began many years ago. Miscellaneous accounts of utilizing wind-powered mills for the purpose of irrigation and grinding by ancient civilizations of Egypt, Babylon, Persia, and China are found in various records (Leung and Yang, 2012). The earliest windmills were found to have a vertical axis. These windmills were simple devices that were used to grind grains.

The first details of horizontal axis windmills are reported from about 1000 AD in historical documents from Tibet, China, and Persia. It was introduced to the Mediterranean region and Europe (Şahin, 2004). Employing windmills to generate electricity was first demonstrated in 1887 by James Blyth in Scotland. These windmills are known as wind turbines, and the electricity is generated by the conversion of kinetic energy into mechanical energy and then to electrical energy. Today, the largest producer of wind energy in the world is China, followed by the United States, Germany, India, and Spain. Indeed, wind energy technologies on a large scale have great potential to boost the world's economy. As an alternative energy source, these can be of great use in the form of wind energy farms.

12.2 Feasibility of Wind Energy Resources

An essential concern for the development of wind energy technologies is the accurate and reliable research of wind energy resources at a specific geological site. This research influences the cost and effectiveness of wind energy production. To establish a long-term and large-scale wind energy farm, the research of wind measurements over past years is a must. The calculation of average wind velocity over time is the most important factor for establishing small- and large-scale wind energy plants at any site.

Accurate calculations, modelling, and designing are required to install wind energy plants at a specific geological site. The wind energy available for conversion mainly depends on the average wind speed and the turbine's swept area. In a typical calculation of the available wind power, the kinetic energy (KE) of wind, a function of mass (m) and velocity (v) is given by

$$KE = \frac{1}{2}mv^2 \tag{12.1}$$

The volume of the wind passing through an area 'A' with velocity 'v' for 't' seconds, is equal to 'Avt'. So, the mass of wind with density 'ρ' will be

$$m = \rho \times V = \rho Avt \tag{12.2}$$

From Equations 12.1 and 12.2, the kinetic energy can be obtained as

$$KE = \frac{1}{2}\rho Atv^3 \tag{12.3}$$

Hence, the total wind power can be calculated as

$$P = \frac{KE}{\text{time}} = \frac{1}{2}\rho A v^3 \tag{12.4}$$

Therefore, the total wind power is directly proportional to the cube of average wind speed, the area swept by wind turbine blades, and the density of the air. Practically, the actual output power 'P_t' obtained from wind turbines is lower than the total power because of energy loss in the whole turbine system. Thus the value of turbine efficiency or power coefficient $\left(C = \dfrac{P_t}{P} \right)$ always remains less than 1, and the actual wind power is given by

$$P_t = \frac{1}{2}C\rho A v^3 \tag{12.5}$$

Hence, an increment in the effective area of turbine blades and wind velocities can increase the output power of any wind energy plant. However, this increment can only be done within the stability limits of the wind energy plant. The excess length of turbine blades and very high wind velocities make the foundation and tower units unstable, causing the failure of the whole wind energy plant (Abkar and Porté-Agel, 2015; Zendehbad et al., 2017; Du et al., 2019; Chou et al., 2019).

Besides the structural stability of the turbine system and wind availability at the site, the transportation facilities and availability of sufficient land are also major points of concern. In hilly areas with good wind resources, the transportation of large turbine blades is difficult and the installation and maintenance cost becomes high. The challenges in proper grid integration possibilities in remote areas and other environmental challenges of the project should also be well reviewed before the installation of the wind energy plant. Thus, on the basis of real on-site observations and measurements, the feasibility of wind energy resources can be evaluated.

12.3 Wind Turbines and Their Classification

The wind turbine is a device that converts wind energy to electrical energy. The wind energy is used to rotate the large blades (wings) of wind turbines that are connected to an electricity generator unit. The axis of rotation of these blades is an important factor in wind turbine technology. Based on the axis of rotation of the blades, wind turbines are classified into two categories: horizontal axis wind turbines (HAWT) and vertical axis wind turbines (VAWT) (Eriksson et al., 2008). A systematic representation of this classification is provided in Figure 12.1.

In a HAWT, the axis of rotation of blades is parallel to the ground and wind direction. These turbines have high efficiency and do not require any external power source to start. With modern technologies, these are stable at high wind speed conditions and have good productivity even at low wind speed conditions (Wright and Wood, 2004). HAWT also has some disadvantages: high installation cost, occupation of large ground area, high maintenance cost, noise pollution, etc. Due to their high efficiency and high power generation, HAWT systems are widely used in large-scale wind energy farms.

In VAWT, the axis of rotation of the rotor is perpendicular with respect to the ground and the direction of the wind. These are relatively easier to build and transport. Since these are mounted closer to the ground, the handling of turbulence in VAWT is better in comparison to the HAWT. The designing and installation of these turbines are easy and cost-effective. Since these require less area for installation and low maintenance cost, these can be used for domestic or small-scale private purposes. The main disadvantage of VAWT is their low efficiency. These require a small powered motor for initial rotation. However, depending upon the geological site and environmental conditions, these can be used in place of HAWT units to minimize transportation, installation, and maintenance cost (Bhutta et al., 2012; Saad and Asmuin, 2014; Johari et al., 2018).

On the basis of geological site preferences, the wind turbines are classified into onshore wind turbines and offshore wind turbines. Onshore wind turbines are a category of turbines that are installed on land.

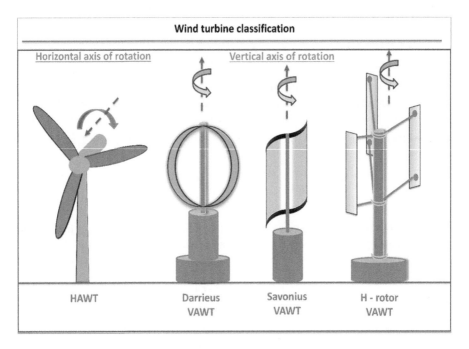

FIGURE 12.1 Classification of wind turbines on the basis of axis of rotation of rotor.

These turbines usually can have a wide range of tower height and rotor diameter. The rotor assembly speed in modern-day onshore turbines varies between 10 and 30 rpm; hence, these can operate even at low wind speeds. The strong foundation on the ground and easy access to various kinds of controllers allow the onshore wind turbines to operate even in high storm conditions.

Offshore wind turbines are installed beyond the coast areas in sea bodies. The significant availability of high wind speed and steady and uniform ocean wind waves has accelerated offshore wind farm's development in the past few years (DeCastro et al., 2019; Jansen et al., 2020; Enevoldsen and Jacobson, 2021). Moreover, most of the industrialized cities of nations are generally located in coastal areas; hence, making offshore wind farms is useful in terms of easy power transmission. The issue of large land occupancy of onshore wind farms can be solved by developing more offshore wind plants. Offshore turbine technology is mostly similar to that of onshore ones. However, the design of the foundations and tower system is different. The foundation of offshore wind turbine system is generally of floating type (in deep waters) or submerged underwater tower type (near the coast) (Figure 12.2) (Roddier et al., 2010; Oh et al., 2018; Esteban et al., 2019; Yang et al., 2019a; Le et al., 2020). There are many advantages of offshore wind farms, such as higher average wind speed, low wind turbulence, low impact of noise pollution on population, etc. On the other hand, higher capital cost, restricted access in poor weather conditions, high maintenance of submarine cables, etc. are some of the disadvantages of offshore wind farms (Dicorato et al., 2011; Röckmann et al., 2017).

12.4 Components of Wind Turbine

The advancement in wind energy harvesting technology directly depends upon the efficiency of the wind turbine unit. In other words, the technologies related to various components of the wind turbine directly affect wind energy harvesting. A conventional wind turbine mainly consists of an anemometer, blades, rotor, gearbox, breaks, generator, and yaw drive units (Hyvärinen and Segalini, 2017). The basic operation of these units are discussed below, and a systematic diagram of these components is provided in Figure 12.3.

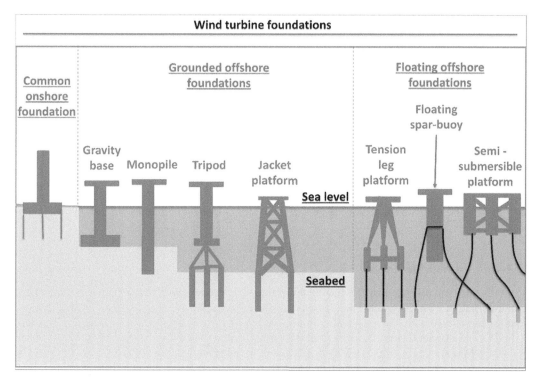

FIGURE 12.2   Schematic representation of various types of wind turbine foundations.

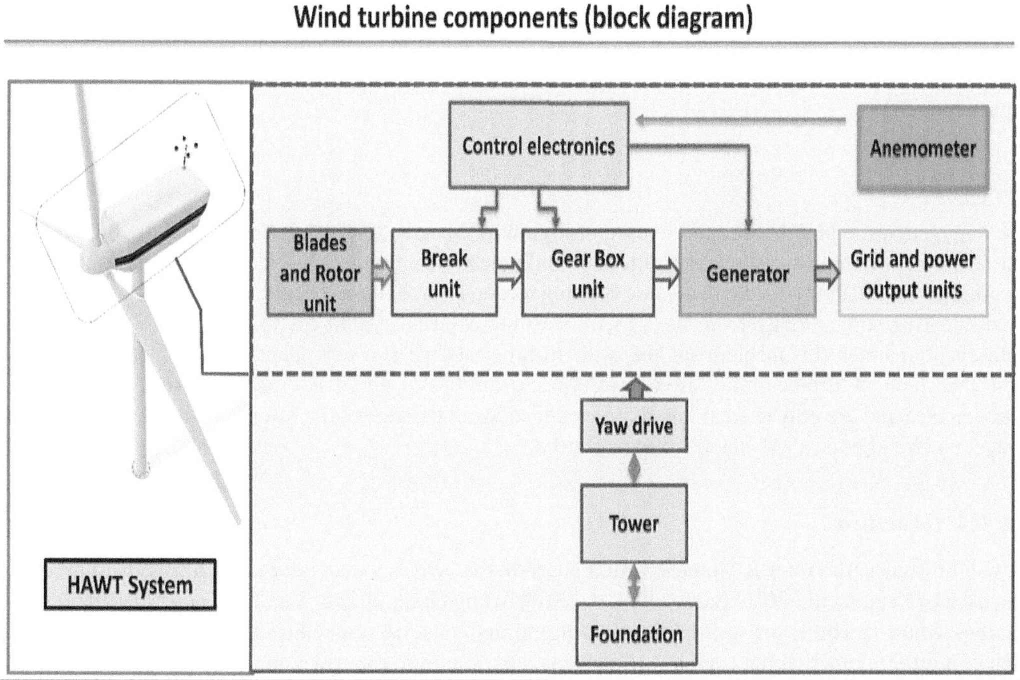

Wind turbine components (block diagram)

FIGURE 12.3   Block diagram of HAWT component system.

12.4.1 Anemometer

An anemometer is a weather station instrument used to measure wind direction and speed. The most common and widely used wind anemometers are cup anemometers and vane anemometers. Cup anemometer has a vertical axis fitted with three or four cups that rotate with the wind flow, and the revolutions per minute of this rotating system are measured to know the wind speed (Ramos-Cenzano et al., 2019). At the same time, a vane anemometer has a horizontal axis of rotation and uses a combined propeller and control tail to detect the wind speed and direction (Thepnurat et al., 2018). The laser and ultrasonic anemometers are relatively new measurement devices that detect the phase shifting of coherent light and sound by the air molecules (Nakai and Shimoyama, 2012; Angelou et al., 2012; Ghahramani et al., 2019). A hot-wire anemometer uses the temperature difference produced between wires to detect the wind speed (Hyvärinen and Segalini, 2017). In practice, depending upon their durability and sensitivity, these anemometers and their specially modified models are used in different geological sites according to the weather conditions and optimization requirements.

12.4.2 Rotor and Blades

A rotor is the main and front part of the wind turbine, which rotates with the help of attached blades upon interaction with the wind. The efficiency of the rotor for various wind speeds limits the rotational speed of blades and hence plays an important role in the overall efficiency of the turbine (Sun et al., 2017). A typical wind turbine of ~5 MW unit capacity has a rotor diameter of up to 130 m. For ~ 8–10 MW units, the rotors can have a diameter up to 150 m. As the energy harvested depends upon the area swept by the rotor system, a smaller rotor becomes more economical for a high wind speed site while a large rotor becomes more economical at low wind speed sites. To increase the efficiency, the modern wind turbine blades are now built from lightweight glass fibre, or carbon fibre-reinforced polymers for high stiffness and strength (Dai and Mishnaevsky, 2015; Cousins et al., 2019; Wang et al., 2020a; Puttaraju and Hanumantharaju, 2020). The proper aerodynamical designing of blades for extracting maximum energy from the wind and stable functioning of the rotor at high wind speeds is necessary. In contrast to the low-power turbines where blades are bolted directly to the hub, the pitch control gearboxes installed in the hub within the rotor of high-power turbines are used to adjust the best blade angles to get maximum power output (Chen et al., 2017; Agarwala, 2019).

12.4.3 Brakes

For emergencies, maintenance, and in case of high wind speeds, the brakes are used in a wind turbine. These are of two main types: electrical brakes and mechanical breaks (Spooner and Williamson, 1996; Sirotkin et al., 2018). An electromagnetic braking torque is used in electrical breaks, which is generated by dissipating some energy from the generator to the external circuit resistors. When operated in an auto-cyclic manner, this mechanism keeps the turbine rotating at a safe speed even in fast wind conditions. Mechanical turbine brakes are of two types: drum brakes and disk brakes. These require regular maintenance and are only used at low rotor speeds to avoid damage to the rotor. To slow the rotor to low speed, electrical brakes and blade furling are used.

12.4.4 Gear Box

In wind turbines, the rotor is coupled with a gearbox that drives a generator at high speeds to produce electricity (Nejad et al., 2016; Salameh et al., 2018; Wang et al., 2020b; Van De Kaa et al., 2020). The gearbox converts slow rotor speed (~50–100 rpm) to high generator speed (~1,000–1,500 rpm). In every step of a turbine functioning, such as turbine start-ups, shut-downs, grid connection transfer, and emergency stops, the turbine rotor applies large force to the turbine drive train; therefore, a turbine gearbox is designed to support such high loads.

12.4.5 Generator

The mechanical power obtained from wind energy is converted into electrical power with the help of a wind turbine generator. These wind turbine generators are slightly different from other common generator units due to the fluctuating torque provided by the turbine rotor. Two types of AC generators are used in large wind turbine systems: synchronous generators and induction generators. However, DC generators are also used in small wind turbine systems (Majeed et al., 2018). Asynchronous generator operates at a constant speed related to the fixed frequency (Ali et al., 2017; Shi et al., 2018; Kumar et al., 2020b; Li et al., 2020c). A DC supply is provided with the help of slip rings and brushes to excite the rotor field windings of synchronous generators. This kind of synchronous generator is also called an electrically excited synchronous generator or wound rotor synchronous generator (WRSG). An alternative way of generating an excitation field is the use of permanent magnets instead of electromagnets. This kind of synchronous generator is called permanent magnet synchronous generators (PMSG). On the other hand, the induction generators do not require brushes and slip ring arrangements for providing DC supply. These are mostly self-excited by shunt capacitors, which draw the excitation power from utility grids. The induction generator is widely used in both small and large wind plants (Zhu et al., 2017; Zhang et al., 2020; Abolvafaei and Ganjefar, 2020).

12.4.6 Foundation and Tower

A tower standing on a site-specific foundation is used to mount the turbine unit at a suitable height. The building materials, strength, height, and design of the tower vary according to the turbine unit's stability at the installation site (Hamed et al., 2020; De Lana et al., 2021). The tower structure stands on a strong foundation. The design and construction of an onshore turbine's foundation are relatively simpler than that of offshore turbine's foundation. While a common steel and concrete foundation is used in onshore wind energy plants, the foundations of offshore wind farms can be of several types. Among several designs of such foundations, the monopile foundations are usually made out of steel bars raised from the bottom of seabed. The gravity foundations are formed at the seabed with the help of concrete-filled with sand and stones. Another offshore foundation type is a tripod steel structure used to fix at the seabed with three smaller piles (Figure 12.2). All these types of foundations are used where water surface level lies up to ~50 m of seabed. For deeper waters, floating type foundations are used.

12.4.7 Yaw drive

A yaw drive is a very useful unit of wind turbine systems. Mounted on the top of the tower, it allows the turbine unit to rotate horizontally and maintain the blades perpendicular to the wind direction to get maximum efficiency (Song et al., 2018; Saenz-Aguirre et al., 2020; Dai et al., 2021).

12.5 Current Status of Wind Energy Technologies

Studies suggest that wind power potential is far greater than the present-day global energy consumption. Due to awareness of global warming in the last few decades, immense growth has been observed in the wind energy sector. The scenario is that wind power plants provide maximum power among all renewable energy resources in most countries. This is because wind power plants require comparably minimum maintenance cost is negligible, and it is also easily available as well as fuel- and pollution free (Brockway et al., 2019).

The development of high-strength, flexible, and lightweight materials such as thermoplastic and ceramic composites, reinforced epoxy functionalized carbon nanotubes, E-glass fiber-reinforced polymer composites, etc. for manufacturing rotor blades has increased significantly in recent years. New coating materials for the protection of turbine blades from rain and wind erosion are also being developed (Elhadi Ibrahim et al., 2020; Singh et al., 2020; Hong et al., 2021; Elhenawy et al., 2021). The efforts are being made to design variable rotor-speed pitch control systems for the turbine to operate at high wind speeds (Colombo et al., 2020; Bundi et al., 2020; Sarkar et al., 2020; Yuan et al., 2020). Various reliable controller algorithms based on fuzzy logic are being introduced for utilization in pitch controller units (Civelek, 2020; Chavero-Navarrete et al., 2020). New methodologies and technological advancements for increasing the efficiency of doubly-fed induction generators (DFIG) have also been reported in recent years (Boubzizi et al., 2018; Mosaad et al., 2019).

Research is also being carried out to develop cross-axis turbines (Chong et al., 2017). Remarkable progress has been made to develop technology to detect and eliminate ice for wind turbines located in regions with freezing temperatures (Wei et al., 2020). Moreover, the model for the integration of small-scale wind turbines with electric vehicle charging units is being developed across the globe (Mehrjerdi and Hemmati, 2020). Today, the use of various sensor technologies such as microwave sensors, brillouin-based fibreoptic sensors, etc. has eased the maintenance and inspection of wind turbine blades (Li et al., 2017; Coscetta et al., 2017; Sessarego et al., 2020). Along with the development in the technology of various turbine components, progress is also being made in high-efficiency hybrid energy plants. The solar-wind energy plants, solar-wind-biomass energy plants, and solar-wind-hydro energy plants are being used to provide electricity in remote areas (Fathabadi, 2017; Abdali et al., 2018; Nyemba et al., 2019; Mehrjerdi, 2020; Li et al., 2020b).

Presently, the modelling, testing, and implication of offshore wind technologies such as hybrid monopole foundations, suction bucket foundations, floating VAWT units, etc. are also major thrust area due to the high power demand (Wang et al., 2017; Guo et al., 2018; Wang et al., 2019; Yang et al., 2019b). The integration of wind turbine power output with the conventional electric grids and their stability has always been a challenge in the field of wind energy technologies. The research in the development of effective methods and electronics such as high-quality electric hubs and active power filters for the smooth integration of wind energy output with the grids is still a demanding area (Perera et al., 2017; Tareen et al., 2017). The energy storage systems compatible with wind energy farms are also being developed (Hemmati, 2017; Ma et al., 2017; Wei et al., 2018). Apart from all these current developments, the establishment of reliable models and software for the smooth functioning of each unit and various components of wind energy farms is a necessity. Therefore, to fulfil the requirement, various software and models for such purposes are being developed and extensively utilized (Zhu et al., 2018; Ebrahimi et al., 2019; Ti et al., 2020; Vijayakumar and Robinson, 2020; Figaj et al., 2020). The existing technologies are being applied and utilized while new technologies will continue to develop simultaneously in upcoming years until we are capable of utilizing maximum wind power.

12.6 Challenges with Wind Energy Systems

Apart from being one of the emerging green energy technologies, wind energy technology faces many challenges. These are accurate understanding of the physics of wind flow at the geological sites of wind power plant operation. Its maintenance, control, and stability of turbine units and other environmental challenges are also of concern. Noise pollution, climate change, deforestation, bird fatality, and disturbance in marine eco-systems are some of the areas of concern that are discussed further.

12.6.1 Noise Pollution

Noise is a major environmental impediment in the development of wind power energy (Ruggiero et al., 2015; Deshmukh et al., 2019; Zagubień and Wolniewicz, 2019). The number of turbines in operation, temperature, humidity, ground surface material, barriers, etc. are the factors contributing to the wind turbine noises. The noise produced from the wind turbines is categorized into two types: mechanical noise and aerodynamic noise. Mechanical noise usually originates from the wind turbine components such as gearbox, generator, hydraulic systems, etc. The noise produced by these mechanical constituents is more narrow banded (20–100 Hz) and tonal in nature, which is more annoying to humans. Mechanical noise can be minimized while designing the turbine, soundproofing the insides of the turbine housing, using anti-vibrational support feet, and using sound insulating curtain (Jianu and Rosen, 2017; Liu, 2017).

Aerodynamic noise is a characteristic swishing sound generated from the air that flows above the rotor blade. It is a more complex and dominant component of the total noise that is produced from the wind turbine. The aerodynamic noise depends upon the size of the turbine, wind speed, and speed of blade rotation. However, the aerodynamic noise is a broadband noise and is composed of different frequencies. This noise can be checked by designing the blades that minimize such noise (Maizi et al., 2018; Liu et al., 2019; Su et al., 2019; Cao et al., 2020).To check the noise level, various measures are being taken into account, like, recommendation of the minimum separation distance between wind farms and nearest habitations and specifying an upper noise limit value that is audible at the nearest dwelling.

12.6.2 Climate Change/Deforestation/Soil Erosion

Major ecological challenges that are faced with developing a wind turbine farm are deforestation, soil erosion, and local climate change (Armstrong et al., 2016; Enevoldsen, 2018; Pryor et al., 2018; Suryakiran et al., 2020; Nazir et al., 2020). Various activities like clearing away forests, excavation of foundations, construction of roads, etc. affect the local environment as well as the flora and fauna of the place. When the plants are removed to settle wind turbine farms, the soil gets exposed to climatic conditions like strong winds and heavy rainfall, resulting in soil erosion. Moreover, the sites of wind energy farms are usually the areas like grasslands, moorlands, and semi-deserts where wind moves unhindered. And these areas have a very fragile ecosystem with very low biodiversity. Construction of wind energy farms with heavy machinery may alter the local ecological balance, leaving the place irreversibly damaged for a long time. During the operation of wind turbines, turbulence is generated in their wake. This turbulence is a small-scale random air movement that alters the vertical mixing of upper and lower levels of the air. This alteration of vertical mixing of air layers causes an impact on the region and surface during both day and night. This results in the randomness in surface temperature. Summing up, the impact of wind turbines on the environment is a debatable topic; therefore, further research and proper optimization have to be carried out, enabling wind energy to be cleaner and more sustainable.

12.6.3 Bird Fatality

Another important issue that is considered while talking about wind turbines is bird mortality (Heuck et al., 2019; Miao et al., 2019; Smallwood and Bell, 2020). Birds get killed on colliding with the rotating propellers of a wind turbine or severely injured on colliding with the turbine towers, nacelles, and various other structures of the wind energy farm. However, an accurate estimation of bird fatality rate is difficult because of factors like variation in the search area and predator removal rates. Factors like arrangement of wind turbines, design of wind turbine, location and layout of the wind farm, species of bird and other climatic variables, etc. play an important role in wind turbine induced bird mortality (Millon et al., 2018; Cabrera-Cruz et al., 2020). Also, the approaching angle between the turbine and bird flight path orientation has shown a significant correlation with collision probability. The construction of wind turbines and their infrastructure causes the destruction of bird's habitat and creates obstruction in their path of roosting locations and natural feeding grounds. The noise generated in the wind turbines also scares away the birds, thereby reducing their territories, which affects foraging behaviour.

12.6.4 Disturbance in the Marine Ecosystem

Marine life is affected by off-shore wind turbines (Madsen et al., 2006; Bergström et al., 2014; Bray et al., 2016; Riefolo et al., 2016). The foundation of the wind turbines and their on-site erection make seawater turbid and block the sunlight. The inclusion of foreign matter on the seabed disturbs and damages the region's flora and fauna, which ultimately leads to the collapse of the marine ecosystem. The distribution of the fishes is also affected by the wind turbines. It is seen that electromagnetic fields and noise around the wind turbines adversely affect the fishes. Maintenance activities, including lubrication and part replacement, cause oil and other waste products to enter into the marine ecosystem and ultimately disrupt it.

12.7 Future Prospectives

Identification and decision of major steps for the future development of wind energy is the first step towards sustainable energy production. Wind power is the renewable energy resource which, along with solving the demands of the global energy problem, is also the source of rural/regional development, employment opportunities, and creating domestic industry (Munday et al., 2011; Ortega-Izquierdo et al., 2020). However, the infrastructures concerning wind energy exploitation should be evaluated for their impact on the environment. In the future, the introduction of larger turbines with faster speed is expected to be introduced. Another important fact is that the properties of wind at any geological site should be thoroughly and carefully studied. With the introduction of new and advanced technologies, the ongoing research is making a strong platform for the development of high-efficiency turbine systems. Moreover, to establish high-efficiency wind energy farms, advanced surveys should be conducted to point out the places where the maximum output of wind energy can be achieved. The establishment and demonstration of small-scale and regional wind turbine systems in urban and rural areas is also important to increase people's awareness of this green energy sector.

In view of structural advancement, carbon fibre rotor blades are expected to be used in the near future since they have the advantage of stability and weight over traditional rotor blades. The research on the development of high-strength composite materials from plastic waste is currently ongoing, which can revolutionize the protection of the environment in the near future. Large wind turbines will definitely be a challenge in the future due to the amount of material and structures utilized in them. Therefore, the development of turbines with a tower height of 10–200 m and a rotor diameter of 120–180 m will be preferred later. It has been found that almost all countries that utilize wind energy for electricity generation have government policies that are specific to wind energy. Thus, it is important for a country to have their wind energy policies well managed.

12.8 Conclusion

Of all the renewable energy options available to humankind, wind energy has been singled out as one of the most promising alternatives for future energy generation. The recent growth of wind power at the global level explains the potential of wind energy. There are several countries that have already met a considerable portion of their domestic energy demand from wind energy, while many other nations have initiated research and development in this field of energy generation. Energy policies and new advances in wind energy technologies determine the pace and expected target of the development of wind power. The focused research in wind energy technologies has always been centred towards increasing the efficiency of various components used in wind turbines. Nowadays, the development in the field of wind energy technologies is mainly governed by research in off shore wind power technologies. However, most of the wind power generation of the world at the present time is onshore cantered. Offshore wind power has attracted attention due to various advantages over onshore generation. With the growth in the electric vehicle industry, small-scale onshore wind turbine systems can be well utilized for power generation. Compared to fossil fuels, wind power is an environmentally cleaner source of energy, yet it still has few adverse effects on the environment as well as on human and animal life. Although the adverse environmental effects of wind energy technologies are very few, yet these effects cannot be ignored, and more advanced technologies must be introduced to mitigate them because the ultimate aim of sustainable development is utilizing the resources without hampering the environment.

REFERENCES

Abdali, A.L.M., B.A. Yakimovich and V.V. Kuvshinov. 2018. Hybrid power generation by using solar and wind energy. *Energy* 2(3): 26–31.

Abkar, M. and F. Porté-Agel. 2015. Influence of atmospheric stability on wind-turbine wakes: A large-eddy simulation study. *Physics of Fluids* 27(3):035104.

Abolvafaei, M. and S. Ganjefar. 2020. Maximum power extraction from fractional order doubly fed induction generator based wind turbines using homotopy singular perturbation method. *International Journal of Electrical Power & Energy Systems* 119:105889.

Agarwala, R. 2019. Development of a robust nonlinear pitch angle controller for a redesigned 5MW wind turbine blade tip. *Energy Reports* 5:136–144.

Ali, R.B., H. Schulte and A. Mami. 2017, May. Modeling and simulation of a small wind turbine system based on PMSG generator. In *2017 Evolving and Adaptive Intelligent Systems (EAIS)* (pp. 1–6). IEEE.

Angelou, N., J. Mann, M. Sjöholm and M. Courtney. 2012. Direct measurement of the spectral transfer function of a laser based anemometer. *Review of Scientific Instruments* 83(3):033111.

Armstrong, A., R.R. Burton, S.E. Lee, et al. 2016. Ground-level climate at a peatland wind farm in Scotland is affected by wind turbine operation. *Environmental Research Letters* 11(4):044024.

Bergström, L., L. Kautsky, T. Malm, R. Rosenberg, M. Wahlberg, N.Å. Capetillo and D. Wilhelmsson. 2014. Effects of offshore wind farms on marine wildlife—a generalized impact assessment. *Environmental Research Letters* 9(3):034012.

Bersalli, G., P. Menanteau and J. El-Methni. 2020. Renewable energy policy effectiveness: A panel data analysis across Europe and Latin America. *Renewable and Sustainable Energy Reviews* 133:110351.

Bhutta, M.M.A., N. Hayat, A.U. Farooq, et al., 2012. Vertical axis wind turbine–A review of various configurations and design techniques. *Renewable and Sustainable Energy Reviews* 16(4):1926–1939.

Boubzizi, S., H. Abid and M. Chaabane. 2018. Comparative study of three types of controllers for DFIG in wind energy conversion system. *Protection and Control of Modern Power Systems* 3(1):1–12.

Bray, L., S. Reizopoulou, E. Voukouvalas, T. Soukissian, et al. 2016. Expected effects of offshore wind farms on Mediterranean marine life. *Journal of Marine Science and Engineering* 4(1):18.

Brockway, P.E., A. Owen, L.I. Brand-Correa and L. Hardt. 2019. Estimation of global final-stage energy-return-on-investment for fossil fuels with comparison to renewable energy sources. *Nature Energy* 4(7):612–621.

Bulut, U. and G. Muratoglu. 2018. Renewable energy in Turkey: Great potential, low but increasing utilization, and an empirical analysis on renewable energy-growth nexus. *Energy Policy* 123:240–250.

Bundi, J.M., X. Ban, D.W. Wekesa and S. Ding. 2020. Pitch control of small H-type Darrieus vertical axis wind turbines using advanced gain scheduling techniques. *Renewable Energy* 161:756–765.

Cabrera-Cruz, S.A., J. Cervantes-Pasqualli, M. Franquesa-Soler, et al. 2020. Estimates of aerial vertebrate mortality at wind farms in a bird migration corridor and bat diversity hotspot. *Global Ecology and Conservation* 22:e00966.

Cao, J.F., W.J. Zhu, W.Z. Shen, J.N. Sørensen and Z.Y. Sun. 2020. Optimizing wind energy conversion efficiency with respect to noise: A study on multi-criteria wind farm layout design. *Renewable Energy* 159:468–485.

Chavero-Navarrete, E., M. Trejo-Perea, J.C. Jáuregui-Correa, et al. 2020. Hierarchical pitch control for small wind turbines based on fuzzy logic and anticipated wind speed measurement. *Applied Sciences* 10(13):4592.

Chen, Z.J., K.A. Stol and B.R. Mace. 2017. Wind turbine blade optimisation with individual pitch and trailing edge flap control. *Renewable Energy* 103:750–765.

Chong, W.T., W.K. Muzammil, K.H. Wong, C.T. Wang, M. Gwani, Y.J. Chu and S.C. Poh. 2017. Cross axis wind turbine: Pushing the limit of wind turbine technology with complementary design. *Applied Energy* 207:78–95.

Chou, J.S., Y.C. Ou and K.Y. Lin. 2019. Collapse mechanism and risk management of wind turbine tower in strong wind. *Journal of Wind Engineering and Industrial Aerodynamics* 193:103962.

Civelek, Z. 2020. Optimization of fuzzy logic (Takagi-Sugeno) blade pitch angle controller in wind turbines by genetic algorithm. *Engineering Science and Technology, An International Journal* 23(1):1–9.

Colombo, L., M.L. Corradini, G. Ippoliti and G. Orlando. 2020. Pitch angle control of a wind turbine operating above the rated wind speed: A sliding mode control approach. *ISA Transactions* 96:95–102.

Coscetta, A., A. Minardo, L. Olivares, et al. 2017. Wind turbine blade monitoring with Brillouin-based fiber-optic sensors. *Journal of Sensors* 2017:1–5.

Cousins, D.S., Y. Suzuki, R.E. Murray, J.R. Samaniuk and A.P. Stebner. 2019. Recycling glass fiber thermoplastic composites from wind turbine blades. *Journal of Cleaner Production* 209:1252–1263.

Dai, G. and L. Mishnaevsky Jr. 2015. Carbon nanotube reinforced hybrid composites: computational modeling of environmental fatigue and usability for wind blades. *Composites Part B: Engineering* 78:349–360.

Dai, J., T. He, M. Li, and X. Long. 2021. Performance study of multi-source driving yaw system for aiding yaw control of wind turbines. *Renewable Energy* 163:154–171.

De Lana, J.A., P.A.A.M. Júnior, C.A. Magalhães, et al. 2021. Behavior study of prestressed concrete wind-turbine tower in circular cross-section. *Engineering Structures* 227:111403.

DeCastro, M., S. Salvador, M. Gómez-Gesteira, et al. 2019. Europe, China and the United States: Three different approaches to the development of offshore wind energy. *Renewable and Sustainable Energy Reviews* 109:55–70.

Deshmukh, S., S. Bhattacharya, A. Jain and A.R. Paul. 2019. Wind turbine noise and its mitigation techniques: A review. *Energy Procedia* 160:633–640.

Dicorato, M., G. Forte, M. Pisani and M. Trovato. 2011. Guidelines for assessment of investment cost for offshore wind generation. *Renewable Energy* 36(8):2043–2051.

Du, W., W. Dong, H. Wang and J. Cao. 2019. Dynamic aggregation of same wind turbine generators in parallel connection for studying oscillation stability of a wind farm. *IEEE Transactions on Power Systems* 34(6):4694–4705.

Ebrahimi, S., M. Jahangiri, H.A. Raiesi and A.R. Ariae. 2019. Optimal planning of on-grid hybrid microgrid for remote Island using HOMER software, Kish in Iran. *International Journal of Energy* 3(2):13–21.

Elhadi Ibrahim, M. and M. Medraj. 2020. Water droplet erosion of wind turbine blades: Mechanics, testing, modeling and future perspectives. *Materials* 13(1):157.

Elhenawy, Y., Y. Fouad, H. Marouani and M. Bassyouni. 2021. Performance analysis of reinforced epoxy functionalized carbon nanotubes composites for vertical axis wind turbine blade. *Polymers* 13(3):422.

Enevoldsen, P. 2018. A socio-technical framework for examining the consequences of deforestation: A case study of wind project development in Northern Europe. *Energy Policy* 115:138–147.

Enevoldsen, P. and M.Z. Jacobson. 2021. Data investigation of installed and output power densities of onshore and offshore wind turbines worldwide. *Energy for Sustainable Development* 60:40–51.

Eriksson, S., H. Bernhoff and M. Leijon. 2008. Evaluation of different turbine concepts for wind power. *Renewable and Sustainable Energy Reviews* 12(5):1419–1434.

Esteban, M.D., J.S. López-Gutiérrez and V. Negro. 2019. Gravity-based foundations in the offshore wind sector. *Journal of Marine Science and Engineering* 7(3):64.

Fathabadi, H. 2017. Novel grid-connected solar/wind powered electric vehicle charging station with vehicle-to-grid technology. *Energy* 132:1–11.

Figaj, R., M. Żołądek and W. Goryl. 2020. Dynamic simulation and energy economic analysis of a household hybrid ground-solar-wind system using TRNSYS software. *Energies* 13(14):3523.

Ghahramani, A., M. Zhu, R.J. Przybyla, et al. 2019. Measuring air speed with a low-power MEMS ultrasonic anemometer via adaptive phase tracking. *IEEE Sensors Journal* 19(18):8136–8145.

Guo, Y., L. Liu, X. Gao and W. Xu. 2018. Aerodynamics and motion performance of the H-type floating vertical axis wind turbine. *Applied Sciences* 8(2):262.

Hamed, Y.S., A.A. Aly, B. Saleh, et al. 2020. Nonlinear structural control analysis of an offshore wind turbine tower system. *Processes* 8(1):22.

Hemmati, R. 2017. Technical and economic analysis of home energy management system incorporating small-scale wind turbine and battery energy storage system. *Journal of Cleaner Production* 159:106–118.

Heuck, C., C. Herrmann, C. Levers, et al. 2019. Wind turbines in high quality habitat cause disproportionate increases in collision mortality of the white-tailed eagle. *Biological Conservation* 236:44–51.

Hong, S., Y. Wu, J. Wu, et al. 2021. Microstructure and cavitation erosion behavior of HVOF sprayed ceramic-metal composite coatings for application in hydro-turbines. *Renewable Energy* 164:1089–1099.

Hyvärinen, A. and A. Segalini. 2017. Effects from complex terrain on wind-turbine performance. *Journal of Energy Resources Technology*, 139(5):051205.

Jansen, M., I. Staffell, L. Kitzing, et al. 2020. Offshore wind competitiveness in mature markets without subsidy. *Nature Energy* 5(8):614–622.

Jianu, O.A. and M.A. Rosen. 2017. Preliminary assessment of noise pollution prevention in wind turbines based on an exergy approach. *European Journal of Sustainable Development Research* 1(2):12.

Johari, M., M. Jalil and M.F.M. Shariff. 2018. Comparison of horizontal axis wind turbine (HAWT) and vertical axis wind turbine (VAWT). *International Journal of Engineering and Technology* 7(4.13):74–80.

Khan, Z., S. Ali, M. Umar, D. Kirikkaleli and Z. Jiao. 2020. Consumption-based carbon emissions and international trade in G7 countries: The role of environmental innovation and renewable energy. *Science of the Total Environment* 730:138945.

Kumar N, N. Jeena and H. Singh. 2019. Elevated temperature modulates rice pollen structure: A study from foothill Himalayan Agro-ecosystem in India. *3Biotech* 9: 175.

Kumar, A., P. Kumar, H. Singh and N. Kumar. 2020a. Adaptation and mitigation potential of roadside trees with bio-extraction of heavy metals under vehicular emissions and their impact on physiological traits during seasonal regimes, Urban Forestry & Urban Greening https://doi.org/10.1016/j.ufug.2020.126900.

Kumar, R.R., P. Devi and C. Chetri, et al. 2020b. Design and characteristics investigation of novel dual stator pseudo-pole five-phase permanent magnet synchronous generator for wind power application. *IEEE Access* 8:175788–175804.

Le, C., J. Zhang, H. Ding, P. Zhang and G. Wang. 2020. Preliminary design of a submerged support structure for floating wind turbines. *Journal of Ocean University of China* 19(6):1265–1282.

Leung, D.Y. and Y. Yang. 2012. Wind energy development and its environmental impact: A review. *Renewable and Sustainable Energy Reviews* 16(1):1031–1039.

Li, H.X., D.J. Edwards, M.R. Hosseini and G.P. Costin. 2020a. A review on renewable energy transition in Australia: An updated depiction. *Journal of Cleaner Production* 242:118475.

Li, J., Liu, P. and Z. Li. 2020b. Optimal design and techno-economic analysis of a solar-wind-biomass off-grid hybrid power system for remote rural electrification: A case study of west China. *Energy* 208:118387.

Li, L., H. Li, M.L. Tseng, H. Feng and A.S. Chiu. 2020c. Renewable energy system on frequency stability control strategy using virtual synchronous generator. *Symmetry* 12(10):1697.

Li, Z., A. Haigh, C. Soutis, A. Gibson, and R. Sloan. 2017. Microwaves sensor for wind turbine blade inspection. *Applied Composite Materials* 24(2):495–512.

Liu, Q., W. Miao, C. Li, W. Hao, H. Zhu and Y. Deng. 2019. Effects of trailing-edge movable flap on aerodynamic performance and noise characteristics of VAWT. *Energy* 189:116271.

Liu, W.Y. 2017. A review on wind turbine noise mechanism and de-noising techniques. *Renewable Energy* 108:311–320.

Ma, Y., W. Cao, L. Yang, F. Wang and L.M. Tolbert. 2017. Virtual synchronous generator control of full converter wind turbines with short-term energy storage. *IEEE Transactions on Industrial Electronics* 64(11):8821–8831.

Madsen, P.T., M. Wahlberg, J. Tougaard, K. Lucke and A.P. Tyack. 2006. Wind turbine underwater noise and marine mammals: Implications of current knowledge and data needs. *Marine Ecology Progress Series* 309:279–295.

Maizi, M., M.H. Mohamed, R. Dizene and M.C. Mihoubi. 2018. Noise reduction of a horizontal wind turbine using different blade shapes. *Renewable Energy* 117:242–256.

Majeed, Y.E., I. Ahmad and D. Habibi. 2018. A multiple-input cascaded DC–DC converter for very small wind turbines. *IEEE Transactions on Industrial Electronics* 66(6):4414–4423.

Majid, M.A. 2020. Renewable energy for sustainable development in India: Current status, future prospects, challenges, employment, and investment opportunities. *Energy, Sustainability and Society* 10(1):1–36.

Mehrjerdi, H. 2020. Modeling and optimization of an island water-energy nexus powered by a hybrid solar-wind renewable system. *Energy* 197:117217.

Mehrjerdi, H. and R. Hemmati. 2020. Stochastic model for electric vehicle charging station integrated with wind energy. *Sustainable Energy Technologies and Assessments* 37:100577.

Miao, R., P.N. Ghosh, M. Khanna, W. Wang, and J. Rong. 2019. Effect of wind turbines on bird abundance: A national scale analysis based on fixed effects models. *Energy Policy* 132, pp.357–366.

Millon, L., C. Colin, F. Brescia and C. Kerbiriou. 2018. Wind turbines impact bat activity, leading to high losses of habitat use in a biodiversity hotspot. *Ecological Engineering* 112:51–54.

Mosaad, M.I., A. Abu-Siada and M.F. El-Naggar. 2019. Application of superconductors to improve the performance of DFIG-based WECS. *IEEE Access* 7:103760–103769.

Munday, M., G. Bristow and R. Cowell. 2011. Wind farms in rural areas: How far do community benefits from wind farms represent a local economic development opportunity? *Journal of Rural Studies* 27(1):1–12.

Nakai, T. and K. Shimoyama. 2012. Ultrasonic anemometer angle of attack errors under turbulent conditions. *Agricultural and Forest Meteorology* 162:14–26.

Nazir, M.S., N. Ali, M. Bilal, and H.M. Iqbal. 2020. Potential environmental impacts of wind energy development: A global perspective. *Current Opinion in Environmental Science & Health* 13:85–90.

Nejad, A.R., Y. Guo, Z. Gao, and T. Moan. 2016. Development of a 5 MW reference gearbox for offshore wind turbines. *Wind Energy* 19(6):1089–1106.

Nyemba, W.R., S. Chinguwa, I. Mushanguri and C. Mbohwa. 2019. Optimization of the design and manufacture of a solar-wind hybrid street light. *Procedia Manufacturing* 35:285–290.

Oh, K.Y., W. Nam, M.S. Ryu, J.Y. Kim and B.I. Epureanu. 2018. A review of foundations of offshore wind energy convertors: Current status and future perspectives. *Renewable and Sustainable Energy Reviews* 88:16–36.

Ortega-Izquierdo, M. and P. del Río. 2020. An analysis of the socioeconomic and environmental benefits of wind energy deployment in Europe. *Renewable Energy* 160:1067–1080.

Perera, A.T.D., V.M. Nik, D. Mauree and J.L. Scartezzini. 2017. Electrical hubs: An effective way to integrate non-dispatchable renewable energy sources with minimum impact to the grid. *Applied Energy* 190:232–248.

Pryor, S.C., R.J. Barthelmie and T.J. Shepherd. 2018. The influence of real-world wind turbine deployments on local to mesoscale climate. *Journal of Geophysical Research: Atmospheres* 123(11):5804–5826.

Puttaraju, D.G. and H.G. Hanumantharaju. 2020, February. Investigation of bending properties on carbon fiber reinforced polymer matrix composites used for micro wind turbine blades. In Journal of Physics: Conference Series (Vol. 1473, No. 1, p. 012049). IOP Publishing.

Ramos-Cenzano, A., M. Ogueta-Gutierrez and S. Pindado. 2019. On cup anemometer performance analysis and improvement: A (still) ongoing process. *Dynamics* 38:41.

Riefolo, L., C. Lanfredi, A. Azzellino, et al. 2016, June. Offshore wind turbines: an overview of the effects on the marine environment. In *The 26th International Ocean and Polar Engineering Conference*. International Society of Offshore and Polar Engineers.

Röckmann, C., S. Lagerveld and J. Stavenuiter. 2017. Operation and maintenance costs of offshore wind farms and potential multi-use platforms in the Dutch North Sea. In Buck, B., Langan, R. (eds.) *Aquaculture Perspective of Multi-use Sites in the Open Ocean* (pp. 97–113). Springer, Cham.

Roddier, D., C. Cermelli, A. Aubault and A. Weinstein. 2010. WindFloat: A floating foundation for offshore wind turbines. *Journal of Renewable and Sustainable Energy* 2(3):033104.

Ruggiero, A., J. Quartieri, C. Guarnaccia and S. Hloch. 2015. Noise pollution analysis of wind turbines in rural areas. *International Journal of Environmental Research* 9(4):1277–1286.

Saad, M.M.M. and N. Asmuin 2014. Comparison of horizontal axis wind turbines and vertical axis wind turbines. *IOSR Journal of Engineering (IOSRJEN)* 4(08):27–30.

Saenz-Aguirre, A., E. Zulueta, U. Fernandez-Gamiz, A. Ulazia and D. Teso-Fz-Betono. 2020. Performance enhancement of the artificial neural network–based reinforcement learning for wind turbine yaw control. *Wind Energy* 23(3):676–690.

Şahin, A.D. 2004. Progress and recent trends in wind energy. *Progress in Energy and Combustion Science* 30(5):501–543.

Salameh, J.P., S. Cauet, E. Etien, A. Sakout and L. Rambault. 2018. Gearbox condition monitoring in wind turbines: A review. *Mechanical Systems and Signal Processing* 111:251–264.

Sarkar, S., B. Fitzgerald, and B. Basu. 2020. Individual blade pitch control of floating offshore wind turbines for load mitigation and power regulation. *IEEE Transactions on Control Systems Technology* 29(1):305–315.

Sessarego, M., J. Feng, N. Ramos-García and S. G. Horcas. 2020. Design optimization of a curved wind turbine blade using neural networks and an aero-elastic vortex method under turbulent inflow. *Renewable Energy* 146:1524–1535.

Shi, K., H. Ye, W. Song and G. Zhou. 2018. Virtual inertia control strategy in microgrid based on virtual synchronous generator technology. *IEEE Access* 6:27949–27957.

Singh, S., H.G. Hanumantharaju, S.R.T. Yadla, S.H.R. Yadav and S. Srivatsa. 2020, February. Investigation of bending properties of E-Glass fiber reinforced polymer matrix composites for applications in micro wind turbine blades. In *Journal of Physics: Conference Series* (Vol. 1473, No. 1, p. 012047). IOP Publishing.

Sirotkin, E.A., E.V. Solomin, S.A. Gandzha and I.M. Kirpichnikova. 2018. Backup mechanical brake system of the wind turbine. In *Journal of Physics: Conference Series* (Vol. 944, No. 1, p. 012109). IOP Publishing.

Smallwood, K.S. and D.A. Bell. 2020. Effects of wind turbine curtailment on bird and bat fatalities. *The Journal of Wildlife Management* 84(4):685–696.

Song, D., J. Yang, X. Fan, et al. 2018. Maximum power extraction for wind turbines through a novel yaw control solution using predicted wind directions. *Energy Conversion and Management* 157:587–599.

Spooner, E. and A.C. Williamson. 1996. Direct coupled, permanent magnet generators for wind turbine applications. *IEE Proceedings-Electric Power Applications* 143(1):1–8.

Stokes, L.C. and C. Warshaw. 2017. Renewable energy policy design and framing influence public support in the United States. *Nature Energy* 2(8):1–6.

Su, J., H. Lei, D. Zhou, et al. 2019. Aerodynamic noise assessment for a vertical axis wind turbine using Improved Delayed Detached Eddy Simulation. *Renewable Energy* 141:559–569.

Sun, Z., M. Sessarego, J. Chen and W.Z. Shen. 2017. Design of the OffWindChina 5 MW wind turbine rotor. *Energies* 10(6):777.

Suryakiran, M.N.S., W. Begum, R.S. Sudhakar and S. K. Tiwari. 2020. Development of wind energy technologies and their impact on environment: A review. In Siano, P., Jamuna, K. (eds.) *Advances in Smart Grid Technology* (pp. 51–62). Springer, Singapore.

Tareen, W.U., S. Mekhilef, M. Seyedmahmoudian and B. Horan. 2017. Active power filter (APF) for mitigation of power quality issues in grid integration of wind and photovoltaic energy conversion system. *Renewable and Sustainable Energy Reviews* 70:635–655.

Thepnurat, M., P. Saphet and A. Tong-On. 2018, December. Low-Cost DIY vane anemometer based on LabVIEW interface for Arduino. In *Journal of Physics: Conference Series* (Vol. 1144, No. 1, p. 012028). IOP Publishing.

Ti, Z., X.W. Deng and H. Yang. 2020. Wake modeling of wind turbines using machine learning. *Applied Energy* 257:114025.

Van de Kaa, G., M. Van Ek, L.M. Kamp and J. Rezaei. 2020. Wind turbine technology battles: Gearbox versus direct drive-opening up the black box of technology characteristics. *Technological Forecasting and Social Change* 153:119933.

Vijayakumar, G. and M. Robinson. 2020. ExaWind: A multifidelity modeling and simulation environment for wind energy. In *Journal of Physics: Conference Series* (Vol. 1452, No. 1, p. 012071). IOP Publishing.

Wang, B., Y. Ming, Y. Zhu, et al. 2020a. Fabrication of continuous carbon fiber mesh for lightning protection of large-scale wind-turbine blade by electron beam cured printing. *Additive Manufacturing* 31:100967.

Wang, S., A. Nejad, E.E. Bachynski and T. Moan. 2020b. A comparative study on the dynamic behaviour of 10 MW conventional and compact gearboxes for offshore wind turbines. *Wind Energy* 24:770–789.

Wang, X., X. Yang and X. Zeng. 2017. Centrifuge modeling of lateral bearing behavior of offshore wind turbine with suction bucket foundation in sand. *Ocean Engineering* 139:140–151.

Wang, X., X. Zeng, X. Li and J. Li. 2019. Investigation on offshore wind turbine with an innovative hybrid monopile foundation: An experimental based study. *Renewable Energy* 132:129–141.

Wang, Y., D. Zhang, Q. Ji and X. Shi. 2020c. Regional renewable energy development in China: a multidimensional assessment. *Renewable and Sustainable Energy Reviews* 124:109797.

Wei, K., Y. Yang, H. Zuo and D. Zhong. 2020. A review on ice detection technology and ice elimination technology for wind turbine. *Wind Energy* 23(3):433–457.

Wei, L., Z. Liu, Y. Zhao, G. Wang and Y. Tao. 2018. Modeling and control of a 600 kW closed hydraulic wind turbine with an energy storage system. *Applied Sciences* 8(8): 1314.

Wright, A.K. and D.H. Wood. 2004. The starting and low wind speed behaviour of a small horizontal axis wind turbine. *Journal of Wind Engineering and Industrial Aerodynamics* 92(14–15):1265–1279.

Yang, W., W. Tian, O. Hvalbye, et al. 2019a. Experimental research for stabilizing offshore floating wind turbines. *Energies* 12(10):1947.

Yang, Y., M. Bashir, C. Li and J. Wang. 2019b. Analysis of seismic behaviour of an offshore wind turbine with a flexible foundation. *Ocean Engineering* 178:215–228.

Yuan, Y., X. Chen and J. Tang. 2020. Multivariable robust blade pitch control design to reject periodic loads on wind turbines. *Renewable Energy* 146:329–341.

Zagubień, A. and K. Wolniewicz. 2019. The impact of supporting tower on wind turbine noise emission. *Applied Acoustics* 155:260–270.

Zendehbad, M., N. Chokaniand and R.S. Abhari. 2017. Measurements of tower deflections on full-scale wind turbines using an opto-mechanical platform. *Journal of Wind Engineering and Industrial Aerodynamics* 168:72–80.

Zhang, J., M. Cui and Y. He. 2020. Robustness and adaptability analysis for equivalent model of doubly fed induction generator wind farm using measured data. *Applied Energy* 261: 114362.

Zhu, J., J. Hu, W. Hung, et al. 2017. Synthetic inertia control strategy for doubly fed induction generator wind turbine generators using lithium-ion supercapacitors. *IEEE Transactions on Energy Conversion* 33(2):773–783.

Zhu, W.J., W.Z. Shen, E. Barlas, F. Bertagnolio and J.N. Sørensen. 2018. Wind turbine noise generation and propagation modeling at DTU Wind Energy: A review. *Renewable and Sustainable Energy Reviews* 88:133–150.

13

Emerging Trends in Hydropower Energy

Rajat Singh, Monika Bisht, and Manendra Singh

CONTENTS

13.1 Introduction

Clash for water is likely to increase in the near future due to population growth and socioeconomic development, and water resources managers will face tough choices when water is distributed to the users. Water is a basic resource used in multiple sectors, including the environment; the appropriation is an inherently social-political process, which is likely to become increasingly analyzed as the competition becomes between the different sectors. Since markets are usually absent or inadequate, the appropriation of water between competing demands is concluded administratively taking into account key, often clashing, objectives such as equity, economic efficiency, and maintenance of ecological integrity. Various allocation instruments have been developed to accommodate the efficiency and equity principles (Dinar et al., 1997; Molle et al., 2007). Another crisis associated with water resources management comes from the fact that many waters using activities generate externalities downstream. Since water flows downstream together with outside the settlement area, the natural spatial scale at which water appropriation decisions can be made is the river basin. Policy equipment is designed to achieve a certain level of equity and economic efficiency, therefore, can be advanced and implemented at scale of quantity of waterflow in river (Davis, 2007).

Water for crop irrigation is usually appropriated by a system of annual rights to use a fixed, static volume of water, which is commonly less than what farmers would forecast (Young, 2005). Farmers' demand for water is determined by the value and usage of water in crop production, which is influenced by crop water requirements and crop prices. In the continuing method, which is one of the most prevailing

DOI: 10.1201/9781003175926-13

valuation methods used in irrigated agriculture, one usually concludes constant crop water requirements. In Turkey, the South-eastern Anatolia Project was planned with a fixed irrigation water demand of about 10,000 m³/ha/yr (Kolars and Mitchell, 1994). Moving from a static to a dynamic allotment process in a fully allocated basin involves that the policies are regularly updated based on the hydrologic status of the river basin. It also gives rise to the development of river basin management methods that increase water productivity. Dynamic management strategies are generally used in the hydropower sector, both regulated and deregulated electricity markets. In regulated electricity markets, an Independent System Operator produces a dispatch based on a least-cost criterion (also called "merit-order" operation); hydropower plants are dispatched to minimize the normal operating costs of the irrigation and hydrothermal electrical system over a given planning season (e.g. 5 years). This activity is regularly updated given the status of the system, which consists of information on the storage levels in the reservoirs and the latest hydrologic information (Pereira, 1989). Again, these agreements are regularly updated as hydrologic conditions, spot prices, and financial position change. In a multipurpose and multi-reservoir system, frequently adjusting release and withdrawal agreements based on the latest hydrologic information will increase the benefits derived from the system. However, the duration for which such an adjustment can be carried out results from complex spatial and temporal interactions between the physical attributes of the water resources system (storage, natural flows), the economic and social consequences of conserving, and the impacts on natural ecosystems.

13.2 Hydropower

Renewable energy plays an important role in using natural resources and help to stimulate toward the clean development mechanism (Nanda, 2016). It helps to achieve the 'Sustainable Development Goal-7' on reliable, clean, and affordable energy (International Renewable Energy Agency, 2012). Across the world, renewable energy is showing extraordinary growth since the last decade. Among the renewable energy, hydropower is also one. The goal of hydropower development is not just to generate electricity, it's also to meet legal requirements, such as the status of the surrounding environment, irrigation area, and habitation area.. The need to provide enough fresh water to a growing population in the face of rising water scarcity and deteriorating water quality has pushed sustainable water resource management to the forefront of the global development agenda. Dams have, for a long time, played an important role in human growth, offering significant social and economic benefits. Around 30%–40% of irrigated land relies on water stored behind dams (World Commission on Dams, 2000), and hydropower contributed for 16% of global electricity in 2008 (IEA, 2010).

The consequences of large hydropower dams can be both positive and negative (Sternberg, 2010). Dams have been built to manage river flows, store water to ensure a steady supply of water during dry periods, control floods, irrigate agricultural lands, facilitate navigation, and generate electricity. Negative effects combined with the construction of massive dams result in population relocation, land degradation, and river flow conversion, impacting water quality downstream (Tilt et al., 2009). Many countries around the world are extending their reliance on hydroelectric dams as a source of electricity and irrigation. However, such creation should be done in a way that considers environmental issues and asks how best to plan water supplies. We use the water footprint principle, which estimates the amount of freshwater absorbed and polluted to generate a service along its supply chain, as an indicator of hydroelectricity's water use. A commodity's water footprint is calculated by dividing the total amount of freshwater absorbed or polluted by the potential of output of the commodity (Hoekstra et al., 2011). The water footprint is made up of three parts: green water (consumptive use of rainwater), blue water (consumptive use of ground or surface water), and grey water (the volume of water polluted). There are 7.6 billion of people around the globe based on the last UNO reports, and this number is constantly rising. This population increases along with the direction of expanding urbanization (over 54%) that appeared in rapid industrialization, especially in the backward regions which require higher energy demands. Naturally, every country goes through various development steps of industrialization until the local flora and fauna have matured to the point where appropriate technology can be used to capture energy in a clean and sustainable manner with the least possible impact on the local flora and fauna. In the 21st century, the rapid pace of local urbanization trends of Latin and South America, African, Asian, and Indian districts eventuate industrialization mechanisms that are many times seen skipping this natural development by implementing the hi-tech systems directly into the individual residence of people who are totally dependent on growing crops. Teaching people how to use Renewable Energy Sources (RES) with 21st-century technology is, thus, a vital issue for discussion and if relevant it needs to be taught step by step. Hydropower is generally known as one of the oldest ways how to harvest nature's energy and capitalize it further (Paish, 2002).

13.3 Types of Hydropower

There are two purposes of water power plants in each particular country, i.e. local and nationwide. In case of local, each accession has some different purpose, tying to fill in the gap of specific mechanical and energy complication that occurs in the given area, e.g. water hammer, funicular, water mill, and local railway for mine and wood. Alternatively, hydropower plants could help to achieve hydrogen and also support local power grids because of the current trends in transport conversion to electric and hybrid power, hence the higher power demands and green energy drawback. It also seen as an applicable option for decentralized generation approach (Table 13.1).

In case of the nationwide one, it has a completely different agenda and is usually based on the country's hydrological system. For example, in Slovakia, most of the hydropower plants are of the fall types, because the bulk of Slovak rivers originate in Slovakia, their aquosity is either low or extremely volatile on the one hand, and high on the other. Taiwanese hydropower plants are a combination of broadly cascading, diversion, and large aggregation types because of the annual precipitation rate, typhoons, and sudden droughts. Furthermore, Taiwan is an island country; thus, it is most important to keep the mountain freshwater on the surface as long as possible.

13.4 Water Efficiency in Agriculture

Globally, agriculture creates the largest user of fresh water, with irrigation departure (abstractions) representing approximately 70% of total water use (Fischer et al., 2007). Of this, an estimated half alone reaches the intended crop—the rest being "lost" around between the point of consideration and the crop. The proportion used in many developing countries is even higher (Turral et al., 2010), highlighting the dependence of rural-based economies on water for agriculture. In Mediterranean Europe, irrigated agriculture accounts for around 60% of total absorption, but production is at risk due to increasing water scarcity and competition also for scarce resources. Climate change threatens to aggravate the situation with an anticipated reduction in water resources (Falloon and Betts, 2010) and a rising in agricultural water demand (Diaz, 2007). In Central and Northern Europe, agricultural absorption is typically very limited (1%–2%) but can still have serious environmental impacts, being fixed in drier catchments in the

TABLE 13.1

Different Type of Hydropower and Capacity

S.No.	Type of Hydropower	Capacity
1.	Large/Very large	>100MW / 0.5GW
2.	Medium	25–100 MW
3.	Small	1–25 MW
4.	Mini	100 kW–1 MW
5.	Micro	5–100 kW
6.	Pico	<5 kW

driest months. Supplemental irrigation here is demanding, especially for horticulture, which gives maximize yield and quality to deliver continuous supplies of vegetables and fruit to the large-scale supermarkets and processors. Distinctly, securing sufficient water for agriculture in the future will be necessary in order to meet the changing food demands of a burgeoning global population. However, agriculture lies at the interface between the environment and society, and any developments in water management need to take into account the multifunctional nature of agriculture and the benefits of diversity it provides, not just to food production.

13.5 Water Footprint for Crop

In the natural world of the Earth, an excessive amount of water is available. Nevertheless, fresh water accounts for 2.5% of overall available water in lakes, wetlands, rivers under the surface and sub-surface as well as groundwater (Rao, 1971). There are many reasons why people use so much water and pollute at such a high rate that it is not sustainable (Moon, 2008). Groundwater levels are dropping, rivers are drying up, and pollution levels are increasing, all signs of increasing water stress (Nace, 1967). The availability of resources, particularly water, has become a debatable topic between policymakers and institutions (Falkenmark and Widstrand, 1992). In areas where water has already been lost or exhausted to critical levels, the severity level is exorbitant (UNESCO, 2003). People use a lot of water for drinking, cooking, and washing but also more for producing goods like food, clothing, paper, cotton, and so on, which is referred to as their "water footprint." After the term "virtual water" was delivered at SOAS, the idea of water footprint was nurtured and maintained (Allan, 1997). "Every year, more water is discharged into the Middle East as 'virtual water' than is discharged down the Nile into Egypt for agriculture" (Allan, 2001). It recognizes the compassion of virtual water and provides a highly functional solution with no known drawbacks. In his writings, Falkenmark (1989) described an experiment in which he classified water as conspicuous (blue water) or non-conspicuous (grey water) depending on how much water was used. Users are conscious of visible water sources, such as streams and rivers, springs, and aquifers. Non-visible water, as well as moisture, is retained in the soil profile and is only added if crops and plants need it (Falkenmark, 1989). Until recently, there has been a notable lack of interest in research and water management activities, as well as studies on water use and emissions. Despite this, the use of manufacturing and supply chains is minimal. As a result, there should be some caution in the case where the organisation and characteristics of a production and supply chain have a significant impact on quantities, both in the temporal and spatial domains. Furthermore, there is a connection between water consumption and pollution and the final consumer product. Allan (2001) coined the term "virtual water," which is described as the amount of water required to produce goods and services. Furthermore, the automatic movement of goods and services from one location to another contributes to water transfer in the latent form. The explanation for this is that water is used in the preparation and production of products and services. For instance, a pair of shoes (bovine leather) requires about 8,000 litres of water to refine and prepare dynamically, while a glass of milk (200 ml) requires about 200 litres of water in latent form to bring into being. Furthermore, 35 and 140 litres of water are needed to prepare and distil a cup of tea (250 ml) and a cup of coffee (125 ml), respectively (Guardian, 2013). Likewise, 1 kg of cereals used in animal feeds necessitates the use of $1.2\,m^3$ of water (Verdegem et al., 2006). Nevertheless, as verified by Allan, the focuses solely on direct use of virtual water. Hoekstra (2003) has recommended a further important explanation as the concept of "water footprint." After the explanation of Chapagain and Hoekstra (2008), it also became known that a clear structure to evaluate the link between human utilization and the provision of the globe is well-established. It can be considered of as freshwater usage indicators that show not just a consumers or producers direct water use, but also the indirect water use across the supply chain. However, the expansion in terms of Blue, Green, and Grey water provides an analytical approach to evaluating direct water use in an efficient manner. It has been tested as a multidimensional indicator that allows for an accurate assessment of water use volumes by source and contaminated volumes by form of pollution; all aspects of a total water footprint can be reported geographically and sequentially. The amount of surface and groundwater consumed by the commodities during processing is indicated by the blue water footprint, which is based on the water footprint standard. In addition, the green water footprint refers to rainwater that has been stored in the soil as soil moisture. The amount of freshwater needed to accept the load of contaminants based on existing ambient water quality requirements is referred to as a commodity's grey water footprint. In this regard, the recently released Water Footprint Assessment Manual-2011 (Hoekstra et al., 2011) provides straight-cut techniques and evaluation approaches. It can be used as a tool to conserve limited domestic water supplies by importing water-intensive goods and exporting goods that do not need as much water. Water-scarce countries, on the other hand, will benefit from shipping water-intensive goods. As a result, agricultural product trade on a national and international scale is increasingly being recognized as a tool for conserving domestic water resources and ensuring national water security.

13.6 Case Study

13.6.1 Indus River Basin

The Indus river basin irrigation channel is one of the largest irrigation channels in the world, covering 16 million hectares and using $123\,km^3$ of water with the help of canals and groundwater for food production. Pakistan mainly depends on the system. It consists of four large hydropower projects, which is namely river dam Ghazi-Barotha (1,450 MW) and three reservoirs Mangla (1,000 MW), Tarbela (3,478 MW), and Chashma (184 MW), developing 11% of total hydropower potential. In Pakistan, irrigation has priority access to water, and energy generation is secondary. According to an analysis of the river basin model, using hydropower concept can increase their economy from USD 2.9 to 3.0 billion, which may be small, but the food production would decrease by the same rate. This can justify the importance of hydropower in the Indus valley river basin.

13.6.2 Aral Sea Basin

This case is known for conflicts between Kyrgyzstan (upstream) and Uzbekistan (downstream) on the Syr Darya River. Energy demand is higher in the winter season for warming and cotton production, while irrigation demand is higher in the summer season. Downstream nations compensate with upstream ones for energy during winter, supplying them gas and oil at subsidized prices. This system was framed at Soviet times but after independence, this upstream–downstream collaboration failed. Upstream regions released water during winter for energy production, which caused a problem for downstream regions and also threatened food and income security; this resulted in conflicts between these two regions and failure of the pact.

13.7 Significance in Indian Context

The total number of dams in India is more than 4,300, and all are in working condition. The main function of these dams is irrigation, and they also contribute to the hydroelectric power generation, flood control, and water supply. At present, to touch the global economic pace, India needs to build additional multipurpose plants (GoI, 2008; World Bank, 2010), and these will help India in adaptation practices with changing climatic condition. Most water resource experts now agree that water disputes are mostly caused by inadequate water management, rather than physical water scarcity. The understanding of water footprint in this context directs improvements in water use capacity, water protection within/between countries, and water loss and surplus areas. In India, there is a large agriculture sector as well as a significant amount of surface and underground water supplies. The fact that India's economy is dependent on agriculture is obvious, and wheat, along with maize, barley, rice, and other grains, is the most important cereal crop. Even so, India is a country where a large portion of the population suffers from hunger and malnutrition. The underutilization of agricultural and water resources as a result of low productivity and unusable farmland may be a major factor for this. Droughts and floods are a common occurrence in India; thus, water resource management is taken into account appropriately. The complexity is linked to insufficient administrative support and programme execution that is weak and inefficient in improving plans based on unpredictable circumstances that can be effectively arranged out of water footprint danger (Falkenmark, 1997; Kumar et al., 2015, 2019). Likewise, aquaculture will naturally reduce the amount of water used in the production of animal feeds compared to feeding cattle, chickens, pigs, and other livestock (Verdegem et al., 2006). At such a critical juncture, issues such as how will potential water demand be met will arise. What is an appropriate average water consumption conclusion? And for how long will it lead to the solution of the water shortage problem? In the current situation, there are just a few pressing issues. The primary concern should be efficient water resource management in order to ensure comprehensiveness.

13.8 Management of Irrigation Water

With the changing climate, it is necessary to take serious management practices especially in irrigation efficiency in the hydropower plant region of India's subcontinent. In this chapter, we explain the best four irrigation practices that farmers must adopt in their agriculture field. Integrating these practices, we can consider these as the best water management practices because these practices have the potential to reduce water loss and soil erosion through meso, micro, small, and large hydropower.

13.8.1 Land Levelling Irrigation

It is defined as the reshaping of the irrigated agriculture land in a systematic way for the efficient and equal distribution of the water in a uniform way through a proper application of the water. This type of practice is generally applied to that area where land is mildly sloped. Previously, this type of irrigation followed the surface method to irrigation of the agriculture field. The slope is the factor that is the effect of this type of irrigation, so for proper irrigation land is levelled for uniform irrigation, which leads to removal of the excess water loss from the field and reduces the erosion through rainfall. This type of irrigation applies to the land having a slope of 8%–12.5%. The advantage of this is to increase the infiltration, reduces surface runoff, and lead to decrease in soil erosion.

13.8.2 Irrigation Water Conveyance

It is the best management practices in which thermoplastic is used underground by replacing of the canal lining. The motive of this thermoplastic pipeline is to increase the water supply, which is lost in the canal lining. It helps to prevent soil erosion and reduce water losses. This irrigation system generally depends upon the pressure of the water, friction losses, flow velocities, and flow capacity. The advantage of the management of these practices is to reduce the water loss by 11% and the disadvantage of it that it diverts the amount of water from the source is decreased (TWDB, 2005).

13.8.3 Irrigation System-Surface Surge Valves

These best management practices are generally implemented to subrogate an on-farm ditch with the help of a gated pipeline to equally distribute water through furrow irrigated fields. This is a type of system in which water is dispersed in a regular interval through a channel through a plugging system, i.e. on and off. The system is included an instrument which has the shape like butterfly valves and it helps to adjust the system in regular interval. This instrument helps to reduce the runoff by increasing the uniformity irrigation and reduce the time of flow to reach the endpoint in the field. The advantage of this system to increase the efficiency of the amount of water delivered to each row and reduce the deep percolation near the head of the water. It helps to save the amount of water by switching to surge flow, estimated to be between 10% and 40% (TWDB, 2005), and it generally depends upon the soil type (structure and texture) and period of the year. This is a tedious type of work in a large watershed to operate the butterfly valves for each and every field. This method is generally not used in the large hydro watershed.

13.8.4 Irrigation Water Management

This type of irrigation system can be managed by the volume, frequency, and application rate of irrigation in a planned, efficient manner that helps to determine the water requirement of the crop, complying with federal, state, and local governmental laws. The volume of water required for irrigation is adjusted on the basis of the seasonal and total rainfall received to date. The frequency or interval of irrigation depends upon the quantity of rainfall, timing, evaporation, etc.

13.9 Conclusion

The status of the river systems and marginal water values are decided on the allocation and distribution of river water to the agriculture field in systematic and dynamic ways. In an area with low rainfall such as arid or dry area, the value of marginal water will be increased and it will decide the priority of the most productive uses. Moreover, to reduce the wastage of water, we should know about the requirement of water for crops and make sure it would be distributed according to the crop requirement. However, in order to ensure efficient water allocation decisions, the community should accept them, and a compensation process should be developed to compensate water users who must reject some of their personnel benefits, so that the chances of public benefits are increased. This principle will help to get the additional benefits obtained through this dynamic and systematic allocation.

REFERENCES

Allan, J.A. 1997. Virtual water: A long term solution for water short Middle Eastern economies? Water Issues Group. School of Oriental and African Studies (SOAS). University of London.

Allan, T. 2001. *The Middle East Water Question: Hydropolitics and the Global Economy*. London: I B Tauris & Co.

Chapagain, A.K. and Hoekstra, A.Y. 2008. The global component of freshwater demand and supply: An assessment of virtual water flows between nations as a result of trade in agricultural and industrial products. *Water International* 33(1): 19–32.

Davis, M. 2007. Integrated water resources management and water sharing. *Journal of Water Resources Planning and Management* 133: 427–445.

Diaz, R.N. 2007. Prescription bottle with time indicator. U.S. Patent Application 29/276,280.

Dinar, A., Rosegrant, M. and Meinzen-Dick, R. 1997. Water allocation mechanisms: Principles and examples, World Bank Technical Paper 1779, Washington, USA.

Falkenmark, M. 1989. The massive water scarcity threatening Africa: Why isn't it being addressed? *Ambio* 18(2): 112–118.

Falkenmark, M. 1997. Meeting water requirement of an expanding world population. *Philosophical Transection, The Royal Society of London* 352: 929–936.

Falkenmark, M. and Widstrand, C. 1992. Population and water resources: A delicate balance. *Population Bulletin (Population Reference Bureau)* 47: 1–36.

Falloon, P. and Betts, R. 2010. Climate impacts on European agriculture and water management in the context of adaptation and mitigation—the importance of an integrated approach. *Science of the Total Environment* 408(23): 5667–5687.

Fischer, E.M., Seneviratne, S.I., Lüthi, D. and Schär, C. 2007. Contribution of land-atmosphere coupling to recent European summer heat waves. *Geophysical Research Letters* 34(6): 1–6.

GoI 2008. Government of India, Ministry of Power: Hydro Power Policy.

Hoekstra, A.Y. 2003. Virtual Water Trade: Proceedings of the International Expert Meeting on Virtual Water Trade. Value of Water Research Report Series, 12. Delft, The Netherlands: UNESCO-IHE.

Hoekstra, A.Y., Chapagain, A.K., Aldaya, M.M. and Mekonnen, M.M. 2011. *The Water Footprint Assessment Manual: Setting the Global Standard*. London: Earthscan.

IEA, 2010. Key world energy statistics 2010, International Energy Agency, Paris, France.

International Renewable Energy Agency. 2012. Hydropower. Renewable energy technologies: Cost analysis series. Abu Dhab, UAE: IRENA.

Kolars, J. and Mitchell, W. 1994. *The Euphrates River and the Southeast Anatolia Project.* Carbondale, IL: Southern Illinois University Press.

Kumar N, Jeena, N. and Singh, H. 2019. Elevated temperature modulates rice pollen structure: A study from foothill Himalayan Agro-ecosystem in India. *3Biotech* 9: 175.

Kumar, N., Kumar, N., Shukla, A., Shankhdhar, S.C. and Shankhdhar, D. 2015. Impact of terminal heat stress on pollen viability and yield attributes of rice (Oryza sativa L.). *Cereal Research Communications* 43(-4): 616–626.

Molle, F., Wester, P., Hirsch, P., Jensena, J., Murray-Rust, H., Paranjpye, V., Pollard, S. and Van der zaag, P. 2007. Water for Food, Water for Life: A Comprehensive Assessment of Water Management in Agriculture, Earthscan and Colombo, International Water Management Institute, London: 585–624.

Moon, B.K. 2008. *Address as Prepared for Delivery to the Davos World Economic Forum.* Switzerland: Davos.

Nace, R. 1967. Are We Running out of Water? Circular No. 536. United States Geological Survey (USGS), Washington DC, USA.

Nanda, N. 2016. Regional cooperation to meet climate pledges of South Asia. *Trade Insights* 12: 18–22.

Paish, O. 2002. Small hydro power: Technology and current status. *Renewable and Sustainable Energy Reviews* 6: 537–556.

Pereira, M. 1989. Optimal stochastic operations of large hydroelectric systems. *Electrical Power and Energy Systems* 11: 161–169.

Rao, K.L. 1971. *India's Water Wealth.* New Delhi: Orient Longman's Ltd.

Sternberg, R. 2010. Hydropower's future, the environment, and global electricity systems. *Renewable and Sustainable Energy Reviews* 14(2): 713–723.

Texas Water Development Board (TWDB), 2005. "Water conservation best management practices (BMP) guide for agriculture in Texas." Based on the Agricultural BMPs contained in Rep. 362, Water Conservation Implementation Task Force, Austin, TX.

Tilt, B., Braun, Y. and He, D. 2009. Social impacts of large dam projects: A comparison of international case studies and implications for best practice. *Journal of Environmental Management* 90(3): S249–S257.

Turral, H., Svendsen, M. and Faures, J.M. 2010. Investing in irrigation: Reviewing the past and looking to the future. *Agricultural Water Management* 97(4): 551–560.

UNESCO, 2003. *World Water Resources at the Beginning of the Twenty-First Century.* (Eds.) I.A. Shiklomanov and J.C. Rodda, Cambridge: Cambridge University Press.

Verdegem, M.C.J., Bosma, R.H. and Verreth, J.A.J. 2006. Reducing water use for animal production through aquaculture. *International Journal of Water Resources Development* 22 (1): 101–113.

World Bank. 2010. Unleashing the Potential of Renewable Energy in India. South Asia Energy Unit, Sustainable Development Department.

World Commission on Dams. 2000. *Dams and Development: A New Framework for Decision-Making.* London: Earthscan.

Young, T.P., D.A. Petersen and J.J. Clary. 2005 The ecology of restoration: Historical links, emerging issues and unexplored realms. *Ecology Letters* 8(6): 662–673.

14

Biochar: Processing Technology, Characterization, and Use in Agriculture

Neha Jeena and Rowndel Khwairakpam

CONTENTS

14.1 Introduction

World energy consumption is presently rising exponentially due to the increase in population. The primary source of energy is fossil fuels. Notwithstanding, based on the environmental effects of CO_2 and worldwide energy issues, fossil fuel replacement has become necessary (Barreto, 2018). Organic waste material, as the major ingredient of solid biomass, has a strong approach for biochar processing. By-products from the cultivation and processing of various agricultural commodities include nutshells, cotton gin, corn cobs, sugarcane bagasse, rice hulls, and straws. The majority of these biomasses are made up of cellulose, hemicellulose, and lignin materials. Besides that, crop residues from forestry, animal manures, municipal waste, food, and other biomass waste materials are suitable for biochar production (Varjani et al., 2019). Biochar is rich in carbon made by heating biomass, such as wood, manure, or leaves, in a sealed container with little or no air. In more scientific terms, biochar is created through the thermal decomposition of organic material in the presence of a limited amount of oxygen (O_2) and at a relatively low temperature (<700°C). The concept "Biochar" is a relatively recent phenomenon, emerging in conjunction with soil management, carbon capture issues, and pollutant immobilization. (Kajitani et al., 2013). Biochar's unique characteristics, including high porosity, high cation exchange capacity, large surface area, functional groups, and stability, make it appropriate for many applications. Biochar has several advantages, including its quick and easy preparation, reusability, eco-friendliness, and low cost (Gayathri et al., 2021). The incorporation of biochar into agricultural soils has been proposed as a way to improve soil fertility while mitigating climate change. Recently, it was reported that converting biomass into biochar can produce not only renewable energy but also reduce CO_2 levels in the atmosphere, implying that more study on biochar impact and behavioural patterns in the soil is needed (Fraser, 2010).

Many factors are driving the growing interest in biochar as a soil amendment today, including sustainable environmental advantages. It is regarded as a unique approach for establishing a substantial and long-term sink for atmospheric CO_2 in terrestrial ecosystems while reducing the need for synthetic fertilizers and generating economic gains through value-added to the agricultural production system and increased crop yields due to improved soil fertility. In reality, recent studies show that producing biochar and applying it to soil can provide instant advantages such as enhanced soil fertility and improved crop production. Biochar has also been used in agricultural fields to remove pollutants from the soil. Biochar becomes a vital ingredient of sustainable agricultural production and soil degradation prevention in a well-managed environment while also providing one of the most realistic ways to combat global warming and stream and groundwater pollution (Spokas et al., 2012). Many studies have been conducted on the pyrolysis of biomass residues for biochar production in which the residues are thermally decomposed at different temperatures. This chapter focuses on providing an overview of processing techniques such as pyrolysis, hydrothermal gasification, torrefaction, characterization through XRD, SEM, TGA, FTIR, BET, etc., and their use in agriculture was discussed.

14.2 Production Techniques of Biochar

The growing interest in using biochar for various applications has resulted in an increase in biomass conversion to biochar. Despite the fact that there are numerous biochar processing methods in the literature, there is no proper classification. Thus, here we highlighted various biochar production methods based on their advances and modernization and categorized them as traditional (i.e. pyrolysis) or modern approaches (i.e. gasification and torrefaction). The biomass type determines the maximum amount of biochar generated, and production conditions like rate of heating, range of temperature, and time of residence all must be adequate. These factors are important because they can influence biochar's physical and chemical states during the manufacturing process (Figure 14.1).

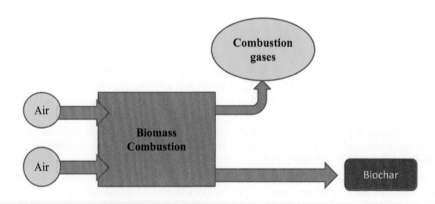

FIGURE 14.1   Basic principles of biochar.

14.2.1 Pyrolysis

Pyrolysis is a thermal degradation process that produces a variety of products such as aqueous, gaseous, and char residues by heating biomass under anaerobic conditions or with a restricted supply of oxygen under the temperature range of 250°C–900°C. In general, agricultural biomass is made up of lignin, cellulose, hemicelluloses, and silica, with cellulose having a pyrolysis point of 350°C and lignin melting above 350°C. This method is a different way of transforming waste biomass into high-value items like biochar, syngas, and bio-oil (Suárez-Abelenda et al., 2017). Depolymerization, fragmentation, and cross-linking of lignocellulosic constituents, including cellulose, hemicellulose, and lignin, occur during the process, yielding solid, liquid, and gaseous products. Biochar is made in various reactors, including paddle kilns, bubbling fluidized beds, wagon reactors, and agitated sand spinning kilns (Wei et al., 2019). The key operating process condition that influences product performance is temperature. As the temperature is raised during the pyrolysis process, the yield of biochar declines, and the production of syngas increases. It is further divided into different forms, such as slow and fast pyrolysis based on process parameters such as temperature, heating rate, and residential time. Slow and fast pyrolysis techniques are considered in traditional methods, while flash pyrolysis, vacuum pyrolysis, and microwave pyrolysis are modern techniques.

14.2.1.1 Slow Pyrolysis

Slow pyrolysis, as the name implies, takes many hours to finish and generates biochar as a primary by-product. It is also known as traditional pyrolysis that involves heating biomass to temperatures between 300°C and 600°C at a heating rate of 5°C–7°C/min. It produces 35%–45% biochar as the main product, 25%–35% bio-oil, and 20%–30% syngas as other products. When compared with other pyrolysis and carbonization methods, the slow pyrolysis innovation produces more biochar. It may be used to increase soil consistency as a dirt enhancer. In this process, a continuous auger/screw pyrolyzer reactor is used. Rice husks, cotton stalks, palmyra nutshell, conocarpus wood wastes, and coconut shells have all been used as a fuel source in the slow pyrolysis process to generate biochar. Notwithstanding this, white paper mill sediment, dairy manure, bull manure, pinewood, and oak wood are also used.

14.2.1.2 Fast Pyrolysis

Fast pyrolysis is a high-efficiency and direct thermochemical method for producing biofuels from biomass. In this procedure, liquefied solid biomass is converted into liquid bio-oil with high potential. It is mostly applied for large-scale biochar production. The approach has two advantages: a short retention period and an increased product recovery rate. However, the main ingredients are bio-oil and syngas rather than biochar. Fast pyrolysis is conducted out in the absence of oxygen at a temperature above 500°C and a heating rate of more than 300°C/min (Wang et al., 2014). The product yield in fast pyrolysis is observed to be bio-oil (60%), biochar (20%), and syngas (20%). To achieve high bio-oil quality, it is necessity to maintain the fume residence time in hot zone which is a differentiating factor of quick pyrolysis technology. It is possible to do this by ensuring that the fumes are quickly extinguished or cooled. Using fast pyrolysis in an Auger pyrolysis reactor at 600°C for 1 minute, researchers created a low-cost catalyst from pyrolysis-derived biochar, which was used for pre-esterification in biodiesel production (Dong et al., 2015). Moreover, Essandoh et al. (2015) observed that pinewood pyrolysis biochar was used to eliminate ibuprofen and salicylic acid from an aqueous solution. Pinewood pyrolysis biochar was made by rapid pyrolysis at 425°C with a 20–30 minutes residence period. Similarly, Laird et al. (2017) developed biochar from hardwood using quick pyrolysis, which had a major effect on crop biomass yield and soil quality.

14.2.2 Gasification

It is a widespread and thermochemical method of producing syngas from solid fuel resources by decomposing them into gaseous products like CO_2, CO, CH_4, H_2, and traces of hydrocarbons. It makes more syngas volume and lower emissions than the other approaches. Furthermore, biochar produced during gasification is considered a waste and has various uses, including dye removal from wastewater, chemical adsorption, carbon sequestration, and as a soil stabilization agent. In this process, reaction temperature (>700°C) is the most important factor in deciding syngas output under a regulated supply of steam and oxygen. It was found that carbon monoxide and hydrogen production are increased as the temperature increased, while carbon dioxide, methane, and hydrocarbons are decreased. Heat transfer within a particle that raises the localized temperature of biomass during the gasification process results in removing water and the gradual release of pyrolytic volatiles. The gasification methodology is divided into two steps :(1) drying, in which the moisture content (depending on the biomass material) of the biomass is entirely evaporated without energy recovery. However, when the biomass contains high moisture content, it is used as a separate process during the gasification process. The next step is (2) oxidation/combustion in which the primary energy sources for the gasification process are oxidation and combustion reactions. These gasification operatives generate CO_2, CO, and water by reacting with the combustible species in the gasifier (Prabakar et al., 2018).

14.2.3 Torrefaction

Torrefaction is a relatively new technique in which biomass is burned at low temperatures (between 230°C and 300°C) to enhance the overall performance of the biomass. It can also be referred to as mild pyrolysis because it uses a low heating rate. On the other hand, it can be thought of as a primary treatment step that enhances the physical, chemical, and biochemical properties of raw biomass, trying to make it more appropriate for combustion, gasification, and co-firing. Various decomposition processes are used to remove oxygen, moisture, and carbon dioxide from the biomass using inert ambient air in the absence of oxygen at a temperature of 300°C. It changes the moisture content, particle size, heating intensity, surface area, energy density, and other properties of biomass. Torrefaction can be accomplished in one of three ways: (1) steam torrefaction: in this method, steam is used to treat the biomass, with a temperature of no more than 260°C and a residence period of about 10 minutes; (2) wet torrefaction, also known as hydrothermal carbonization, occurs when biomass is in contact with water at a temperature of 180°C–260°C for 5–240 minutes; and (3) oxidative torrefaction is a method in which biomass is treated with oxidizing agents such as gases that are used in the combustion process to generate heat energy. This heat energy is used to create the required temperature (Huang et al., 2017; Yu et al., 2017).

14.2.4 Hydrothermal Carbonization

Hydrothermal carbonization is a cost-effective process for producing biochar as it can be done at a low temperature of about 180°C–250°C. Feedstocks with a lot of moisture, like compost, sewage sludge, and animal waste, are turned into biochar using this method. To differentiate the product generated by hydrothermal processes from that produced by dry processes, such as pyrolysis and gasification, the hydrochar is used. There is no compelling reason to dry the biomass until preparing on the grounds that the wet biomass blend is warmed to temperatures ranging from 220°C to 240°C over many hours in a high-pressure (2–10 MPa) reactor (Kambo and Dutta, 2015). To retain equilibrium, the temperature is gradually increased. Rotary drums, kilns, and stoves are used in this process. The majority of organic processes are either dissolved or converted into brown coal. The hydrolyzed product undergoes a series of reactions, including dehydration, fragmentation, and isomerization, to yield the intermediate product, 5-hydroxymethylfurfural, and variants. In addition, the hydrochar is generated through condensation, polymerization, and intramolecular dehydration during the reaction. The retention of nutrients such as N and P, which are beneficial to soil fertility, is an important feature of hydrothermal carbonization-produced biochar. It has many advantages over torrefaction or pyrolysis for biochar processing, including a lower O/C ratio, higher calorific value, stronger grind ability, and enhanced hydrophobicity. The mechanism is complicated by high molecular weight and the dynamic existence of lignin. Dealkylation and hydrolysis reactions initiate lignin decomposition, resulting in phenolic products such as phenols, catechols, and syringols. Repolymerization and cross-linking of intermediates provide the biochar at the end. Like pyrolysis, the lignin components that are not dissolved in the liquid phase are converted into hydrochar (Jain et al., 2016).

14.3 Characterization of Biochar

Biochar characterization is done to see if it can eliminate toxins or be used for other purposes. To establish the relationship among nature and operational conditions with the physical and chemical properties, assess the suitability towards the target applications, and analyze the existence of pollutants and eco-toxicology properties, thorough characterization of biochar is essential prior to any applications. Surface functional groups and structure and elemental analyses are used to characterise biochar with the help of both physical and chemical processes (Figure 14.2).

14.3.1 Physical Characterization

The physical characteristics of biochars play a role in their ability to improve soil quality. Their physical features may have a direct or indirect impact on the soil ecosystem. As biochar is mixed into soils, it can have a major impact on the physical nature of the environment, affecting soil quality, structure, porosity, and tilth by altering aggregate size distribution, bulk density, and other factors (Kambo and Dutta, 2015). Biochar can be characterized by the scanning electron microscopy, one of the physical methods for assessing the macroporosity and physical morphology of solids. The macroporous composition of biochar derived from cellulose plant material is determined by its innate structure, which is critical to the water-holding and pollutants-adsorption capacity of the soil ecosystem. Apart from that, it can also be characterized by other physical methods like X-ray absorption near edge structure, moisture content, ash content, bulk density, and pore volume (Fryda and Visser, 2015).

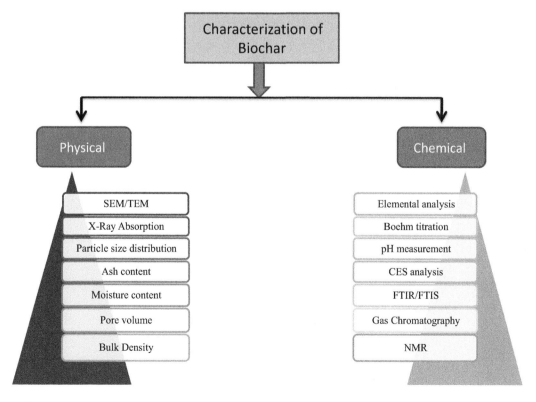

FIGURE 14.2 Characterization of biochar.

14.3.2 Chemical Characterization

The chemical properties of the raw material used to make biochar differ spatially and temporally. Changes in the concentration of elements, including C, H, O, S, and N, are used to evaluate biochar production. It is very well known that solid combustible carbon remains after the sample has been carbonized and the volatile matter has been expelled. Thus, the carbonaceous substances in the solid sample must be analyzed as O/C and H/C ratios that determine the degree of aromaticity and maturation. Apart from that, new advanced methods, like the Fourier Transform Infrared Spectrometer (FTIR), can be used to characterise biochar. It is a vibrational method for examining the surface functional groups of biochar; X-Ray Diffraction (XRD) is a widely applicable method for determining the crystalline and structural properties of biochar; Thermogravimetric Analysis (TGA) is used to observe the physical and chemical properties of materials as a result of temperature increase; Nuclear Magnetic Resonance (NMR) can determine the structural composition of biochar; Brunauer Emmett Teller (BET) helps assess surface area of biochar; Raman spectroscopy can be used for the biochar characterization; etc. (Zhang et al., 2013; Brewer et al., 2014).

14.4 Use of Biochar in Agriculture

See Figure 14.3.

14.4.1 Contribution to the Combating Climate Change

Excess carbon dioxide is released into the atmosphere as a result of the combustion of fossil fuels and the decomposition of biomass, which raises the carbon levels in the atmosphere. Nevertheless, incorporating biochar into the soils can reduce carbon dioxide emissions because biochar can sequester 50% of the carbon from the feedstock. Biochar is highly stable, allowing it to reduce the emission of carbon dioxide from organic decomposition significantly. It also plays an important role in regulating the levels of methane and nitrogen dioxide from the soil (Woolf et al., 2010; Lehmann et al., 2010; Bruckman et al., 2015).

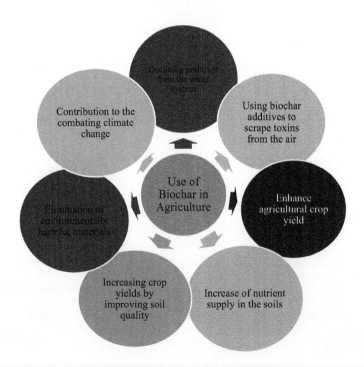

FIGURE 14.3   Use of biochar in agriculture system.

14.4.2 Declining Pollution from the Water System

The use of biochar in the soil significantly reduces offsite pollution. It improves the retention of nutrients such as phosphorus and nitrogen in soils, assists in the reduction of soil nutrient leaching into groundwater, and aids in the preservation of nutrients from erosion due to surface, water flow. As nutrient loss in the soil is reduced, nutrients available for crop cultivation increase which lowers the amount of fertilizer needed for crop growth. There has not been a separate study on erosion yet, but nutrients that are soluble in soil are less likely to be eroded than nutrients that are bound or adsorbed on the surface of soil sediments. Pyrolysis of animal manures may result in a major decline in the accessibility of phosphorous in animal manures. It can also convert soluble inorganic phosphate in manure to adsorbed phosphate in biochar (Blanco-Canqui, 2019; Boehm et al., 2020).

14.4.3 Using Biochar Additives to Scrape Toxins from the Air

Pyrolysis appears to provide additional benefits because incorporation of biochar into the soil results in the removal of pollutants such as sulphur oxides, carbon dioxide, and nitrogen oxides from flue gas, thereby limiting their emissions. Carbon dioxide is precipitated on the surfaces of biochar during an exothermic phase. Such a method could reduce carbon emissions into the atmosphere as a result of fossil fuel combustion. Concurrently, the precipitation process can produce biochar, which is high in nitrogen content and could be used as a replacement for nitrogen-based fertilizers (Martin et al., 2015; Ibrahim et al., 2017; Yang et al., 2020).

14.4.4 Elimination of Environmentally Harmful Materials

Biochar is very effective because it can absorb large environmental pollutants in the soil. Biochar is being used to accumulate many organic contaminants that will potentially change their impact on the atmosphere. Biochar serves as a crucial binding step for various organic contaminants in the ecosystem due to its resistance to microorganisms and exceptional adsorptive ability. There are carbonized and non-carbonized types of organic matter in biochar that play a significant role in adsorption, as these carbonized and non-carbonized portions of biochar adsorb contaminants based on their bulk and surfaces (Xiao et al., 2011; Shen et al., 2019; Khan et al., 2020).

14.4.5 Increasing Crop Yields by Improving Soil Quality

Any process that involves the processing of bioenergy has a detrimental impact on the land and results in unnecessary storage and elimination of biomass. These highly challenging practices have the ability to damage soils unnecessarily, have detrimental effects on soil fertility, destroy habitat, and pollute the environment off-site. Such issues can be rectified using biochar generated by pyrolysis in association with organic matter, as it allows to the recovery of nearly half of the original carbon. Furthermore, biochar is thought to be highly successful in the maintenance of soil fertility (Ding et al., 2016). The exceptional characteristics and advantages of using biochar are not restricted to the region that was disturbed for extracting biomass to produce bioenergy, but it also has the potential to survive in soils for more than 2 years. This demonstrates that applying biochar to lands that are not used for bioenergy production can improve soil productivity while also helping to reduce inorganic chemical contamination of that land's soil (Rhodes, 2012; Zhang et al., 2016).

14.4.6 Increase of Nutrient Supply in the Soils

Biochar can improve the soil texture, porosity, structure, density, and particle size distribution following integration into the soils. Since biochar has a higher porosity and surface area, it will assist in the provision of space for advantageous microorganisms in the soil and aid in the binding of important anions and cations and increase cation exchange ability. The application of biochar raises the pH of the soil, which improves the supply of phosphorus and potassium. As biochar is added to soil, it causes an oxidation process on the surface of the particles. The oxidation of aromatic carbon resulting the formation of carboxyl groups and leads to high cation exchange ability, thereby increasing soil fertility and reduction of leaching in the soil. Nevertheless, a Negative charge is produced on the surface when highly oxidized organic matter attaches to it. Consequently, the positive charge on the sites decreases (Laird et al., 2010; Karimi et al., 2019).

14.4.7 Enhance Agricultural Crop Yield

Biochar causes improved soil fertility, increased seed growth, and ultimately increased agricultural productivity. The significant impact is attributed to the fact that biochar produced by pyrolysis contains more nutrients than the biomass used to make it. Biochar application improves the soil's physical, chemical, and biological properties, which has an indirect effect. Biochar has the primary aim to improve crop response for increased production and productivity (Clough et al., 2013; Yuan et al., 2016). Multiple field trials and pot studies revealed that applying biochar to soils at various rates improved the development and growth parameters of plants such as wheat, maize, beans, cucumber, watermelon, tomato, strawberry, and sweet pepper. This enhancement entails using the modification as a fertilizer carrier or as a pre-loaded nutrient carrier, such as biochar co-compost. Biochar has a wide surface area and is extremely porous, which gives it sorption and retention properties. Rice plant biochar improved rice production and productivity while also retaining nitrogen in waterlogged paddy fields. Amendment of biochar increased crop production and efficiency by reducing drought, salinity, and heat stress effects during the plant development stage and having a significant impact on soil parameters, enrichment and proliferation of beneficial soil microbes, and enhancement of microbial functioning and populations. Biochar increased inorganic nitrogen fertilizer uptake in rice (*Oryza sativa*) plants. Long-term, the continuous addition of charcoal to rice growing soil will increase economic yield and nitrogen fertilizer usage efficiency. At the same time, it will not affect total nitrogen uptake (Fryda and Visser, 2015; Kumar et al., 2017; Kumar et al., 2019).

14.5 Conclusion

In this chapter, the biochar production methods and their applications have been systematically discussed. Biochar is used in several cultivated fields as a source of nutrients, adsorbent, and a fertilizer agent. Biochar addition could also help eradicate plant diseases. The method of processing and type of feedstock utilized has a significant effect on the properties of biochar and its uses. Furthermore, the consistency and effectiveness of biochar in affecting soil quality and plant growth is highly dependent on experimental conditions such as pyrolysis temperature, feedstock content, and biochar age. Field efficiency and economic viability of biochar implementations should be evaluated while offering consistency to the many positive advantages that are a crucial to the problems need to be resolved by further studies.

REFERENCES

Barreto, R.A. 2018. Fossil fuels, alternative energy and economic growth. *Economic Modelling*, 75:196–220.

Blanco-Canqui, H. 2019. Biochar and water quality. *Journal of Environmental Quality*, 48(1):2–15.

Boehm, A.B., C.D. Bell, N.J. Fitzgerald, E. Gallo, C.P. Higgins, T.S. Hogue, R.G. Luthy, A.C. Portmann, B.A. Ulrich and J.M. Wolfand. 2020. Biochar-augmented biofilters to improve pollutant removal from stormwater–can they improve receiving water quality? *Environmental Science: Water Research & Technology* 6(6):1520–1537.

Brewer, C.E., V.J. Chuang, C.A. Masiello, H. Gonnermann, X. Gao, B. Dugan, L.E. Driver, P. Panzacchi, K. Zygourakis and C.A. Davies. 2014. New approaches to measuring biochar density and porosity. *Biomass and Bioenergy* 66:176–185.

Bruckman, V.J., T. Terada, B.B. Uzun, E. Apaydin-Varol and J. Liu. 2015. Biochar for climate change mitigation: tracing the in-situ priming effect on a forest site. *Energy Procedia* 76:381–387.

Clough, T.J., L.M. Condron, C. Kammann and C. Muller. 2013. A review of biochar and soil nitrogen dynamics. *Agronomy* 3:275–293.

Ding, Y., Y. Liu, S. Liu, Z. Li, X. Tan, X. Huang, G. Zeng, L. Zhou and B. Zheng. 2016. Biochar to improve soil fertility. A review. *Agronomy for Sustainable Development* 36(2):1–18.

Dong, T., D. Gao, C. Miao, X. Yu, C. Degan, M. Garcia-Pérez, B. Rasco, S.S. Sablani and S. Chen, 2015. Two-step microalgal biodiesel production using acidic catalyst generated from pyrolysis-derived bio-char. *Energy Conversion and Management* 105:1389–1396.

Essandoh, M., B. Kunwar, C.U. Pittman Jr., D. Mohan and T. Mlsna. 2015. Sorptive removal of salicylic acid and ibuprofen from aqueous solutions using pine wood fast pyrolysis biochar. *Chemical Engineering Journal* 265:219–227.

Fraser, B. 2010. High-tech charcoal fights climate change. *Environmental Science & Technology* 44 (2): 548–549.

Fryda, L., and R. Visser. 2015. Biochar for soil improvement: Evaluation of biochar from gasification and slow pyrolysis. *Agriculture* 5:1076–1115.

Gayathri, R., K.P. Gopinath and P.S. Kumar. 2021. Adsorptive separation of toxic metals from aquatic environment using agro waste biochar: Application in electroplating industrial wastewater. *Chemosphere* 262:128031.

Huang, Y.F., P.H. Cheng, P.T. Chiueh and S.L. Lo. 2017. Leucaena biochar produced by microwave torrefaction: Fuel properties and energy efficiency. *Applied Energy* 204:1018–1025.

Ibrahim, M., G. Li, S. Khan, Q. Chi, and Y. Xu. 2017. Biochars mitigate greenhouse gas emissions and bioaccumulation of potentially toxic elements and arsenic speciation in Phaseolus vulgaris L. *Environmental Science and Pollution Research* 24(24):19524–19534.

Jain, A., R. Balasubramanian and M.P. Srinivasan. 2016. Hydrothermal conversion of biomass waste to activated carbon with high porosity: A review. *Chemical Engineering Journal* 283:789–805.

Kajitani, S., L.H. Tay, S. Zhang and Z.C. Li. 2013. Mechanisms and kinetic modelling of steam gasification of brown coal in the presence of volatile-char interactions. *Fuel* 103: 7–13.

Kambo, H.S. and A. Dutta. 2015. A comparative review of biochar and hydrochar in terms of production, physico-chemical properties and applications. *Renewable and Sustainable Energy Reviews* 45:359–378.

Karimi, A., A. Moezzi, M. Chorom and N. Enayatizamir. 2019. Application of biochar changed the status of nutrients and biological activity in a calcareous soil. *Journal of Soil Science and Plant Nutrition* 20:1–10.

Khan, A.Z., S. Khan, M.A. Khan, M. Alam and T. Ayaz. 2020. Biochar reduced the uptake of toxic heavy metals and their associated health risk via rice (Oryza sativa L.) grown in Cr-Mn mine contaminated soils. *Environmental Technology and Innovation* 17:100590.

Kumar N, N. Jeena, and H. Singh 2019. Elevated temperature modulates rice pollen structure: A study from foothill Himalayan Agro-ecosystem in India. *3Biotech* 9: 175.

Kumar, N., D.C. Suyal, I.P. Sharma, A. Verma and H. Singh 2017. Elucidating stress proteins in rice (Oryza sativa L.) genotype under elevated temperature: A proteomic approach to understand heat stress response. *3 Biotech* 7:205.

Laird, D., P. Fleming, B. Wang, R. Horton and D. Karlen. 2010. Biochar impact on nutrient leaching from a Midwestern agricultural soil. *Geoderma* 158(3–4):436–442.

Laird, D.A., J.M. Novak, H.P. Collins, J.A. Ippolito, D.L. Karlen, R.D. Lentz, K.R. Sistani, K. Spokas and R.S. Van Pelt. 2017. Multi-year and multi-location soil quality and crop biomass yield responses to hardwood fast pyrolysis biochar. *Geoderma* 289:46–53.

Lehmann, J., J.E. Amonette, K. Roberts, D. Hillel and C. Rosenzweig. 2010. Role of biochar in mitigation of climate change. In *Handbook of Climate Change and Agroecosystems: Impacts, Adaptation, and Mitigation*. Imperial College Press, London, pp. 343–363.

Martin, S.L., M.L. Clarke, M. Othman, S.J. Ramsden and H.M. West. 2015. Biochar-mediated reductions in greenhouse gas emissions from soil amended with anaerobic digestates. *Biomass and Bioenergy* 79:39–49.

Prabakar, D., V.T. Manimudi, S. Sampath, D.M. Mahapatra, K. Rajendran and A. Pugazhendhi. 2018. Advanced biohydrogen production using pretreated industrial waste: Outlook and prospects. *Renewable and Sustainable Energy Reviews* 96:306–324.

Rhodes, C.J. 2012. Biochar, and its potential contribution to improving soil quality and carbon capture. *Science Progress* 95(3):330–340.

Shen, Z., Y. Zhang, O. McMillan, D. O'Connor and D. Hou. 2019. The use of biochar for sustainable treatment of contaminated soils. Sustainable Remediation of Contaminated Soil and Groundwater: Materials, Processes, and Assessment, p.119.

Spokas, K.A., K.B. Cantrell, J.M. Novak, D.W. Archer, J.A. Ippolito, H.P. Collins, A.A. Boateng, I.M. Lima, M.C. Lamb, A.J. McAloon and R.D. Lentz. 2012. Biochar: A synthesis of its agronomic impact beyond carbon sequestration. *Journal of Environmental Quality* 41(4):973–989.

Suárez-Abelenda, M., J. Kaal and A.V. McBeath. 2017. Translating analytical pyrolysis fingerprints to Thermal Stability Indices (TSI) to improve biochar characterization by pyrolysis-GC-MS. *Biomass and Bioenergy* 98:306–320.

Varjani, S., G. Kumar and E.R. Rene. 2019. Developments in biochar application for pesticide remediation: Current knowledge and future research directions. *Journal of Environmental Management* 232: 505–513.

Wang, Y., R. Yin and R. Liu. 2014. Characterization of biochar from fast pyrolysis and its effect on chemical properties of the tea garden soil. *Journal of Analytical and Applied Pyrolysis* 110: 375–381.

Wei, J., C. Tu, G. Yuan, Y. Liu, D. Bi, L. Xiao, J. Lu, B.K. Theng, H. Wang, L. Zhang and X. Zhang. 2019. Assessing the effect of pyrolysis temperature on the molecular properties and copper sorption capacity of a halophyte biochar. *Environmental Pollution* 251: 56–65.

Woolf, D., J.E. Amonette, F.A. Street-Perrott, J. Lehmann and S. Joseph. 2010. Sustainable biochar to mitigate global climate change. *Nature Communications* 1(1):1–9.

Xiao, R., J. Bai, Q. Wang, H. Gao, L. Huang, and X. Liu. 2011. Assessment of heavy metal contamination of wetland soils from a typical aquatic–terrestrial ecotone in Haihe River Basin, North China. *CLEAN–Soil, Air, Water* 39(7):612–618.

Yang, W., G. Feng, D. Miles, L. Gao, Y. Jia, C. Li and Z. Qu. 2020. Impact of biochar on greenhouse gas emissions and soil carbon sequestration in corn grown under drip irrigation with mulching. *Science of the Total Environment* 729:138752.

Yu, K.L., B.F. Lau, P.L. Show, H.C. Ong, T.C. Ling, W.H. Chen, E.P. Ng, and J.S. Chang. 2017. Recent developments on algal biochar production and characterization. *Bioresource Technology* 246:2–11.

Yuan, H., T. Lu, Y. Wang, Y. Chen, T. Lei. 2016. Sewage sludge biochar: nutrient composition and its effect on the leaching of soil nutrients. *Geoderma* 267:17–23

Zhang, M., B. Gao, S. Varnoosfaderani, A. Hebard, Y. Yao, and M. Inyang. 2013. Preparation and characterization of a novel magnetic biochar for arsenic removal. *Bioresource Technology* 130:457–462.

Zhang, Y., O.J. Idowu and C.E. Brewer. 2016. Using agricultural residue biochar to improve soil quality of desert soils. *Agriculture* 6(1):10.

15

Socioeconomic Importance of Renewable Energy Sources and Their Impacts on the Environment

G.S. Yurembam and D. Jhajharia

CONTENTS

15.1 Introduction

The supply of continuous energy has played a key role in the functioning of the economy, especially in a developing economy. Long-term energy security in a sustainable way is a viable option for employment generation and to meet the demands of the millennium, which will add up to a country's economic growth. Access to clean energy is also an important pillar for human well-being, economic development, and poverty alleviation.

The global energy production was 14,421 Mtoe (Million tonnes of oil equivalent) in 2018 as compared to 2017 (13959.53 Mtoe), which shows an increase of 3.2%. It was mostly driven by fossil fuels: natural gas, coal, and oil, increasing together by more than 370 Mtoe in 2018 (World Energy Balances: overview, IEA). All renewables and nuclear also increased by 60 and 19 Mtoe, respectively. Fossil fuels ultimately accounted for more than 81% of production in 2018, as was the case in 2017 (IEA, 2020). The cheap cost of electricity generation from these sources in the past led to the dominance of the world's energy supply for these sources. However, energy generation from conventional energy sources leads to uncontrolled emission of CO_2, which is not environment friendly and leads to endangering the livelihood of future generations. Most of the energy of the developing countries is derived from the non-renewable source, of which the resources are decreasing.

The prerequisite for the socio-economic development of a region is the accessibility to a clean and sustainable source of energy. Renewable energy sources such as wind, solar, hydro, and geothermal energy are the primary source for sustainable energy production. Renewable energy (RE) defined as the "energy obtained from the continuous or repetitive currents of energy recurring in the natural environment", encompasses a wide variety of energy sources including solar, wind, biomass, geothermal, hydropower, ocean energy, biofuels, and hydrogen (Moselle, Padilla and Schmalensee, 2010). Renewable energy is considered sustainable because of its environmentally friendly nature and low emissions of harmful

DOI: 10.1201/9781003175926-15

gases. It also benefits society in all dimensions, including economic, social, and environmental. With such an insight, the renewable source of energy holds a promising solution.

15.2 The Theory of Sustainable Development

The notion of a sustainable environment (Brown, 1982) and the management of renewable and non-renewable resources are at the root of the concept of sustainable development (SD). The concept was defined as "development that meets the needs of the present without compromising the ability of future generations to meet their own needs" (WCED, 1987; Bojoet et al., 1992). Traditionally, the three-pillar model has been used to describe sustainability: Ecology, Economy, and Society are all considered to be interconnected and relevant for sustainability (BMU, 1998). The United Nations General Assembly wants to see steps taken to facilitate the convergence of the three pillars of sustainable development—economic development, social development, and environmental protection—as interconnected and mutually reinforcing pillars.

During the round of discussions on the IPCC (Intergovernmental Panel on Climate Change) process, after the first assessment report, which focused on the technologies and cost-effectiveness of mitigation strategies, SD has changed. The Second Assessment Report included concerns of equity and environmental protection (Hourcade et al., 2001). The Third Assessment Report (IPCC, 2007b) addressed global sustainability in depth, and the Fourth Assessment Report (AR4) included chapters on SD in both Working Group (WG) II and III reports, with an emphasis on a review of both climate-first and development-first literature (IPCC, 2007a, b). Energy generation from renewable sources and its applications are essential to meet the Millennium Development Goals. Positive action such as the replacement of fossil fuels with Renewable Energy (RE) sources can bring a set of actions to fulfil the three pillars of sustainable development simultaneously.

With such highlights, this chapter will be emphasized on renewable energy sources and their impacts on social, economic, and environmental aspects. Thus, the goal of this chapter is to summarise and consolidate the material based on the various sources of RE sources and their impacts. The discussion focuses on the impacts of renewable energy sources on economic aspects in Section 15.3.1 and on social aspect in Section 15.3.2. The environmental aspects are represented in Section 15.3.3. Further, Sections 15.3.3.1 and 15.3.3.2 discuss on the aspects of greenhouse gas (GHG) Emissions and Improved human health from renewable energy technologies. Section 15.4 summarizes the overall impact of the transition to renewable energy from the conventional sources of energy.

15.3 Social, Economic, and Environmental Impacts of Renewable Energies

15.3.1 Economic Benefits: Global and Rural Aspect

A number of maximum economic advantages can be reaped from renewable energy sources. The cost of renewable energy is rapidly declining, especially for solar power, which is a major benefit from the RE sources. There is a strong relationship between economic growth and the rise of energy consumption. The energy sector has been considered as a key factor to economic development, and the trend of electricity generation from renewable sources has been recording a changing scenario. Renewable energy stood second in the world electricity generation mix since 2013 and reached almost 26% by the year 2018, which can be seen from Figure 15.1. Power generation from coal was still dominant by far in 2018, reaching 38% of the electricity produced globally. Its share started decreasing in 2018 after having slightly re-increased in 2017, interrupting 4 years of consecutive decrease. Renewables come second in the electricity mix, as has been the case since 2013, and reached almost 26% of the mix in 2018. Although the electricity supply is still dominated by the supply sources of fossil fuels, the data (Lazard) over the last decades show that the scenario has changed as we compared with the renewable energy sources. Figure 15.2 also shows the falling prices of renewable energy sources as compared to other fossil fuels. The change in solar and wind energy prices in the last 10 years can be seen in Figure 15.2.

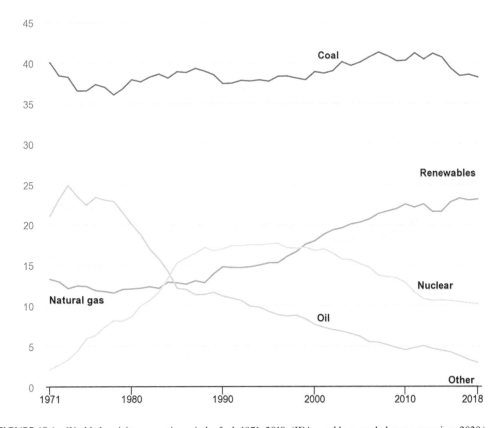

FIGURE 15.1 World electricity generation mix by fuel, 1971–2018. (IEA, world energy balances: overview, 2020.)

Electricity from utility-scale solar photovoltaics cost $359 per MWh in 2009. The price of electricity from solar declined by 89% and onshore wind electricity declined by 70% in just one decade, and the relative price flipped (Roser, 2020). The dip in energy prices from renewable sources also means that the real income of people rises.

Electricity generation from renewable sources such as hydro, solar, biomass, and geothermal can be a promising solution to the increasing demand of energy for driving the production and economic growth ahead. According to a report from IRENA (2017), it was found that reducing global carbon dioxide emissions in line with the Paris Agreement would boost GDP by 0.8% in 2050. The transition to renewable energy will also help in local economic value creation at a regional level. Electricity in remote areas is critical because it can support existing economic growth while also attracting new potential sectors for development. Rural areas attract a large portion of the investment in RE deployment. A good integration of affordable RE technology in the rural sector also provides the opportunity to generate energy that will create new revenue sources, job, and business opportunities. Such a generation of reliable and cheaper electricity can enhance economic development.

15.3.2 Social Benefits

Social aspects are the primary considerations for the progress of any country. Improvement in economic activity will result in social improvement and also alleviate poverty. Renewable energy resources provide a wide opportunity of social benefits like the local employment generation, which will help in the reduction of poverty, improve health condition, provide opportunity for local business, etc.

According to the 2018 Annual review of the IRENA, in 2017, global renewable energy employment reached 10.3 million, up 5.3% from the previous year's figure. The global scenario of employment generation from renewable sources for various countries is presented in Figure 15.3. As per the IRENA

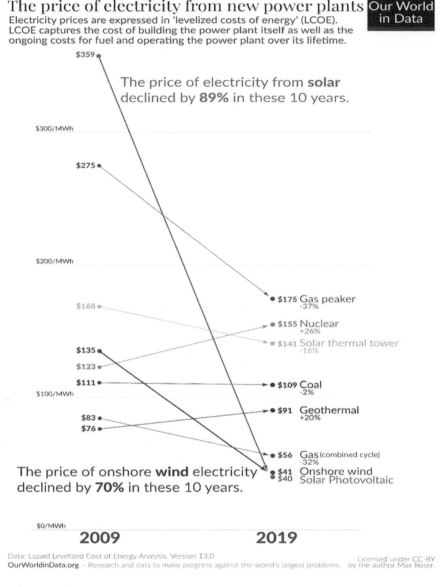

The price of electricity from new power plants

Our World in Data

Electricity prices are expressed in 'levelized costs of energy' (LCOE). LCOE captures the cost of building the power plant itself as well as the ongoing costs for fuel and operating the power plant over its lifetime.

The price of electricity from **solar** declined by **89%** in these 10 years.

The price of onshore **wind** electricity declined by **70%** in these 10 years.

FIGURE 15.2   Costs of electricity generation. (Our World in Data.)

estimates, the expected increase in human welfare from the deployment of renewables is close to 4%, far exceeding the 0.8% rate of improvement in GDP.

China employed the highest number of people (3,880 million jobs), which accounts for almost 43% of the global. In India, the employment from Solar PV touched 164,400 jobs (36% improved), of which 92,400 represented on-grid use (Majid, 2020). Jobs in renewables energy can be categorized into technological development, installation/de-installation, operation, and maintenance. There is an ample amount of opportunity for an increase in the regional-scale business, and RE sectors will prosper with the change as the income of the local population will increase significantly. Employment generated from this RE sector will enable an almost fixed amount of income, better healthcare, and absorption of both skilled and unskilled workers in the job sector. Employment is directly linked to an insured income, and thereby for the well-being of people, the creation and sustainability of jobs are of critical importance in any measure of socio-economic development of a region. The benefits resulting from an increase in savings from

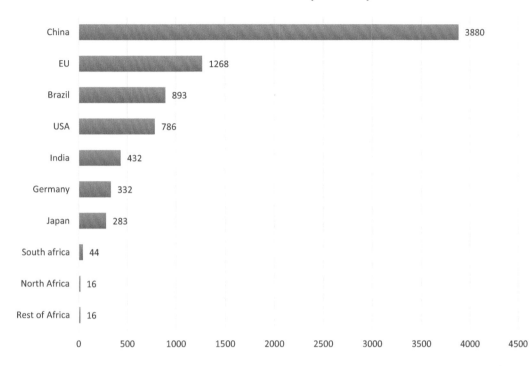

FIGURE 15.3 Employment generation from renewable sources for various countries. (Adapted from Kumar. J and Majid Energy, Sustainability and Society.)

reduced health expenses and environmental externalities overcome the expenses of energy transition that are not reflected in conventional accounting systems.

15.3.3 Environmental Benefits

Almost 79% of the total global energy production comes collectively from conventional sources such as coal, oil, and gas, which have a serious negative effect on GHG emissions (Figure 15.4). The impact of renewable energy on environmental aspects can be seen mostly on the GHG emissions and associated health conditions. The different forms of the contribution of CO_2 gas emissions from different forms of power generation are shown in Figure 15.4. The burning of fossil fuels in a very large quantity to meet the ever-increasing demands of electricity to fuel the growing economy is harmful to human health and are a major source of contributors to climate change. A World Health Organization study has estimated that the increase in greenhouse gases since 1990 has resulted in around 150,000 excess deaths in 2000. Almost all of these people died in countries that are not part of the Organization for Economic Cooperation and Development (OECD), where increased risk factors for malnutrition, diarrhoea, malaria, floods, and cardiovascular disease are attributed to climate change (McMichael et al., 2003).

In recent decades, there has been a visible transition from conventional sources into low-carbon electricity production using renewable energy sources. Renewable energy technologies contribute to mitigating climate change and, when locally produced, decrease the national dependence on imported energy and increase (local) employment.

The focus of the world's attention on environmental issues in recent years has stimulated response in many countries, which has led to a closer examination of energy conservation strategies for conventional fossil fuels (Omer, 2010).

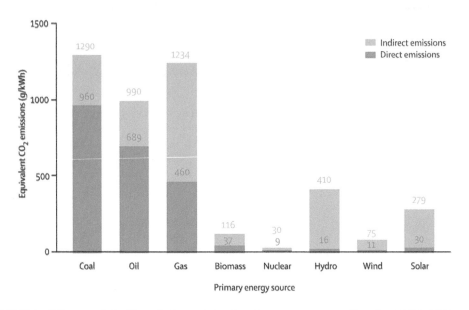

FIGURE 15.4 Full energy chain CO_2 equivalent emissions by primary energy source. (Data from IAEA, 2001.)

15.3.3.1 GHG Emission

According to a report from IRENA (2015), in 2012, a total of 3.1 gigatons (Gt) of CO_2-equivalent (CO_2-e) emissions were avoided globally through the use of renewable electricity (primarily hydropower); total global emissions figure would have been increased by 20% if the power generation was not based on renewable sources. The Union of Concerned Scientists (a US-based organization) did a comparison between conventional forms of power as well as renewable sources. Their analysis reveals that natural gas emits between 0.6 and 2 pounds of carbon dioxide equivalent per kilowatt-hour (CO_2 e/kWh), while coal emits between 1.4 and 3.6 pounds of CO_2 e/kWh when burned for electricity. Wind, on the other hand, emits just 0.02–0.04 pounds of CO_2 equivalent per kWh over its lifetime, compared to 0.07–0.2 for solar, 0.1–0.2 for geothermal, and 0.1–0.5 for hydroelectric (Barriers to Renewable Energy Technologies, 2017). Renewable energy technologies produce lower emissions of CO_2 and other GHGs in comparison to the fossil fuel-based generation of electricity. The various sources of electricity generation are projected to emit substantially less GHG emissions (CO_2e) over their lifetimes than traditional and natural gas plants.

15.3.3.2 Improved Human Health

The most severe consequences of energy on human health are those caused by air pollution from fossil fuel and biomass combustion (Ezzati et al., 2004; Paul et al., 2007). Renewable energy sources produce little or no air pollution. Coal and natural gas plant emissions have been linked to a variety of health issues, including respiratory disorders, cancer, heart attacks, premature death, neurological damage, and a slew of other serious issues. A study conducted by Harvard Medical Centre (2011) found out that life cycle costs and public health effects from coal costs the US public up to 500 billion USD per year, in which many of these costs are from public health and waste management. The health impacts from important air pollutants are also presented in Table 15.1.

Air pollution is a major important issue in many developing countries. In such countries, up to 2.9 billion people still rely on wood, coal, and charcoal for cooking and other domestic purposes. Shifting to cleaner technology, including biomass and solar technologies, can play a role in this regard. Solar and wind energy require essentially no water for operation and maintenance, thereby not polluting the water resources that may be depleted or strained as a result of competition with irrigation, drinking water, or other critical water needs. Among the various sources of renewable energy, biomass and geothermal

TABLE 15.1

Health Impacts of Important Air Pollutants

Primary Pollutants[a]	Secondary Pollutants[b]	Impacts
Particles (PM10, PM2.5, black carbon)		Cardiopulmonary morbidity (cerebrovascular and respiratory hospital admissions, heart failure, chronic bronchitis, upper and lower respiratory symptoms, aggravation of asthma), mortality
SO_2	Sulphates	Like particles[c]
NO_x	Nitrates	Morbidity, like particles[c]
NO_x+VOC	Ozone	Respiratory morbidity, mortality
CO		Cardiovascular morbidity, mortality
Polyaromatic hydrocarbon		Cancers
Lead and mercury		Morbidity (neurotoxic and other)

Adapted from Bickel and Friedrich (2005).

Notes: [a]Emitted by pollution source, [b]created by chemical reactions in the atmosphere, [c]lack of specific evidence, as most available epidemiological studies are based on mass-PM without distinction of components or characteristics (Renewable Electricity Futures Study, 2012).

emit air pollutants but at a rate that is generally much lower than most fossil fuels. Hydroelectric power plants may be present both upstream and downstream of the dam; the river habitats can thus be harmed. According to a study by National Renewable Energy Laboratory's (NREL) explores the implications and challenges of a very high renewable electricity generation from 30% up to 90%, focusing on 80% of all U.S. electricity generation by 2050, which also included biomass and geothermal, noticed that with high renewables, overall water usage and withdrawal would decrease dramatically in the future.

15.4 Conclusion

This chapter highlighted the potential of renewable energy sources as an alternative to the conventional sources of energy with a clean renewable source. It is also clear that the transition towards a low-carbon emission technology for electricity generation has both pros and cons. As the changes towards a more sustainable energy generation, the global renewable energy workforce will continue to expand. Electricity generation from renewable sources such as hydro, solar, biomass, and geothermal can be a promising solution to the increasing demand for energy for driving the production and economic growth ahead. The shifting towards a more renewable source of energy production would boost GDP by 0.8% in 2050. Rural economic value creation will help existing economic growth attract more investment to be developed in the sector. Such a generation of reliable and cheaper electricity can enhance economic development. Improvement in economic activity will also result in social improvement and alleviate poverty. Large opportunities for social benefits like the local employment generation which will help in the reduction of poverty, improvement of health condition, the opportunity for local business, etc, are also noted. Employment is directly linked to an insured income, and thereby improved well-being of people; the creation and sustainability of jobs is of critical importance in any measure of socioeconomic development of a region. Renewable energy technologies also produce lower emissions of CO_2 and other GHGs in comparison to the fossil fuel-based generation of electricity. GHG emissions (CO_2e) are estimated to be substantially lower over the life cycle of various types of electricity production from renewables than from the conventional and natural gas plants. As for human health, renewable energy sources produce little or no air pollution. The multidimensional benefits of renewable energy sources are gaining importance among the decision-makers. The advantages of the transition towards the clean source of energy are numerous, such as in employment, the welfare of society and local industrial growth, and its environmental benefits, all of which enhance its overall acceptability in global and regional societies. An integrated and prioritized approach and policies and models that support renewable development as an alternative sustainable development are promising solutions to harness the benefits of renewable technologies.

REFERENCES

Assembly, U. G. 2005. World Summit Outcome: resolution/adopted by the General Assembly, A/RES/60/1. http://www. refworld. org/docid/44 168a910. Html (October 24, 2005).

Barriers to Renewable Energy Technologies. 2017, Dec 20. Union of Concerned Scientists (UCS).https://www. ucsusa.org/resources/barriers-renewable-energy-technologies.

Bickel, P. and R. Friedrich. 2005. ExternE: Externalities of energy. Methodology, 270.

BMU. 1998. Naturschutz und Re- aktorsicherheit (BMU); Gemeinsames Ministerialblatt, No.1, 10 (in German).

Bojo, J., K. Maler and L. Unemo. 1992. *Environment and Development: An Economic Approach*. Kluwer Academic Publishers, Dordrecht.

Brown, L. R. 1982. Building a sustainable society. *Society* 19(2):75–85.

Ezzati, M., A. D. Lopez, A. A. Rodgers and C. J. Murray. 2004. Comparative quantification of health risks: Global and regional burden of disease attributable to selected major risk factors. World Health Organization.

Hourcade, J. C., P. Shukla, L. Cifuentes, D. Davis, J. Edmonds, B. Fisher and Z. Zhang. 2001. Global, regional, and national costs and ancillary benefits of mitigation. In *Climate change 2001: Mitigation. Contribution of Working Group III to the Third Assessment Report of the Intergovernmental Panel on Climate Change* (pp. 499–559). Cambridge University Press.

International Atomic Energy Agency. 2001. *Sustainable Development Andnuclear Power*. Vienna: IAEA.

IPCC. 2007a. Climate Change 2007: The Physical Science Basis. *Working Group 1 Contribution to the Fourth Assessment Report of the Intergovernmental Panel on Climate Change (IPCC). Technical Summary and Chapter 10*. Solomon, S., Qin, D., Manning, M., Chen, Z., Marquis, M., Averyt, K., Tignor, M.M.B. & Miller, H.L. (eds.), Global Climate Projections.

IPCC. 2007b. Climate Change 2007: Mitigation. *Contribution of Working Group III to the Fourth Assessment Report of the Intergovernmental Panel on Climate Change, 2007.* Metz, B., Davidson, O.R., Bosch, P.R., Dave, R., and Meyer, L.A. (eds.), Cambridge University Press, 851 pp.

IEA (International Energy Agency). 2020. World energy balances: overview. https://www.iea.org/reports/world-energy-balances-overview.

IRENA. 2015. Renewable Energy and Jobs – Annual Review 2015, Abu Dhabi.

IRENA. 2017. Renewable Energy Benefits: Understanding The Socio-Economics.

Majid, M. A. 2020. Renewable energy for sustainable development in India: Current status, future prospects, challenges, employment, and investment opportunities. *Energy, Sustainability and Society* 10(1): 1–36.

McMichael, A. J., D. H. Campbell-Lendrum, C. F. Corvalán, K. L. Ebi, A. Githeko, J. D. Scheraga and A. Woodward. 2003. *Climate Change and Human Health: Risks and Responses*. Geneva: World Health Organization.

Moselle, B., J. Padilla and R. Schmalensee. 2010. *Harnessing Renewable Energy in Electric Power Systems: Theory, Practice, Policy*. Milton Park: Routledge.

Omer, A. M. 2010. Environmental and socio-economic aspects of possible development in renewable energy use. *Journal of Agricultural Extension and Rural Development* 2(1): 001–021.

Paul, W., R. S. Kirk, J. Michael and H. Andrew 2007. A global perspective on energy: health effects and injustices. *Lancet* 370(9591): 965–978.

Renewable Electricity Futures Study. 2012. Energy Analysis. https://www.nrel.gov/analysis/re-futures.html.

Roser, M. 2020.Why did renewable become so cheap so fast? And what can we do to use this global opportunity for green growth? https://ourworldindata.org/cheap-renewables-growth.

WCED, S. W. S. 1987. World commission on environment and development. *Our Common Future* 17(1): 1–91.

Index

For Product Safety Concerns and Information please contact our EU representative GPSR@taylorandfrancis.com Taylor & Francis Verlag GmbH, Kaufingerstraße 24, 80331 München, Germany

Printed and bound by CPI Group (UK) Ltd, Croydon, CR0 4YY

01/05/2025

01858475-0001